Heike Knörzer

Designing, modeling and evaluation of intercropping systems in China

Heike Knörzer

Designing, modeling and evaluation of intercropping systems in China

Improving cropping strategies on the basis of multi-level interactions

Südwestdeutscher Verlag für Hochschulschriften

Imprint
Any brand names and product names mentioned in this book are subject to trademark, brand or patent protection and are trademarks or registered trademarks of their respective holders. The use of brand names, product names, common names, trade names, product descriptions etc. even without a particular marking in this work is in no way to be construed to mean that such names may be regarded as unrestricted in respect of trademark and brand protection legislation and could thus be used by anyone.

Publisher:
Südwestdeutscher Verlag für Hochschulschriften
is a trademark of
Dodo Books Indian Ocean Ltd., member of the OmniScriptum S.R.L Publishing group
str. A.Russo 15, of. 61, Chisinau-2068, Republic of Moldova Europe
Printed at: see last page
ISBN: 978-3-8381-2578-7

Zugl. / Approved by: Stuttgart, Universität Hohenheim, Diss., 2010

Copyright © Heike Knörzer
Copyright © 2011 Dodo Books Indian Ocean Ltd., member of the OmniScriptum S.R.L Publishing group

Table of contents

		Page
	Abbreviations and acronyms	II
	List of figures	VII
	List of tables	VII
1	**Introduction** THE PARADIGM OF SUSTAINABILITY ECOLOGICAL AWARENESS AND RESEARCH DECISION SUPPORT SYSTEM OUTLINE AND HYPOTHESES	1
2	**Field Experiments conducted for publications** EXPERIMENT I: FIELD TRIALS IN CHINA 2008 TO 2009 EXPERIMENT II: FIELD TRIALS IN GERMANY 2007 TO 2009	13
3	**Publications**	20
4	**Chapter I:** The rediscovery of intercropping in China: a traditional cropping system for future Chinese agriculture	21
5	**Chapter II:** A modeling approach to simulate effects of intercropping and interspecific competition in arable crops	57
6	**Chapter III:** Extension and evaluation of intercropping field trials using spatial models	87
7	**Chapter IV / Excursus:** Model-based approach to quantify and regionalize peanut production in the major peanut production provinces in the People's Republic of China	106
8	**Chapter V:** Integrating a simple intercropping algorithm into CERES-wheat and CERES-maize with particular regard to a changing microclimate within a relay-intercropping system	112
9	**Chapter VI:** Evaluation and performance of the APSIM crop growth model for German winter wheat, maize and fieldpea varieties within monocropping and intercropping systems	143
10	**General Discussion** THE FUTURE OF INTERCROPPING? THE FUTURE OF MODELING INTERCROPPING?	167
11	**Summary** Zusammenfassung	177 180
12	**References**	184
	Acknowledgement / Danksagung	200

Abbreviations and acronyms

A	Agressivity
A.D.	Anno domini
AIC	Akaike information criterion
ALMANAC	Agricultural Land Management Alternative with Numerical Assessment Criteria
ANOVA	Analysis of variance
APSIM	Agricultural Production Systems Simulator
AR(1)	First-order autoregressive model
ASA	American Society of Agronomy
BBCH	Biologische Bundesanstalt für Land- und Forstwissenschaft, Bundessortenamt und Chemische Industrie
B.C.	Before christ
C_3	Plants with C_3 carbon fixation
C_4	Plants with C_4 carbon fixation
Ca	Calcium
CERES	Crop-Environment Resource Synthesis
cm	Centimetre
CO_2	Carbon dioxide
cohort1retranslocationwt	Change in retranslocation cohort 1 dry matter
CR	Nutrient competitive ratio
CSIRO	Commonwealth Scientific and Industrial Research Organisation
CSSA	Crop Science Society of America
DAS	Days after sowing
DFG	Deutsche Forschungsgemeinschaft
dlt_dm	Actual above-ground dry matter
dlt_lai_pot	Potential change in live plant leaf area index
dlt_lai_stressed	Potential change in leaf area index allowing for stress
dlt_leaf_no	Change in number of leaves
dlt_leaf_no_pot	Potential leaf number
dm	Dry matter
DSSAT	Decision Support System for Agrotechnology Transfer

Abbreviations and acronyms

e	East plot border
e.g.	For example
ep	Plant water uptake
esw_layr(1, 2, 3)	Extractable soil water in different soil layers
et al.	Et alii (and others)
etc.	Et cetera
Exp	Exponential model
FAO	Food and Agriculture Organization of the United Nations
FASSET	Farm Assessment Tool
Fe	Iron
Fig.	Figure
g	Gram
G1	Kernel number per unit weight at anthesis
G2	Kernel weight under optimum conditions
G2	Potential kernel number
G3	Potential kernel growth rate
G3	Standard stem and spike dry weight at maturity
GAPS	General-purpose simulation model of the Atmosphere-Plant-Soil system
Gau	Gaussian model
GDP	Gross domestic product
grain_n	Nitrogen in grain
grain_no	Grain number
grain_size	Size of each grain
grain_wt	Weight of grain
H	Plant height
H^+	Hydron
ha	Hectare
IBSNAT	International Benchmark Sites Network for Agrotechnology Transfer
i.e.	That is
IE	Internal efficiency of nitrogen use
inter	Intercropping
INTERCOM	Ecophysiological model for crop-weed interactions
IRTG	International Research Training Group

Jr.	Junior
K	Potassium
kg	Kilogram
KJ	Kilo joule
KMS	Simplified Kubelka-Munk equations model
K_2O	Potassium oxide
LAI	Leaf area index
leaf_no_sen	Number of senesced leaves per square meter
LER	Land equivalent ratio
LV	Linear variance model
m	Metre
m^2	Square metre
MJ	Mega joule
mm	Millimeter
Mn	Manganese
MOE	Ministry of Education of the People's Republic of China
mono	Monocropping
N	North
N	Nitrogen
N_2	Molecular nitrogen in the atmosphere
NCP	North China Plain
n_demand	Nitrogen demand of plant
n_demanded(1, 2)	Nitrogen demand of plant
NE	Northeast
n.m.	Not measured
N_{min}	Soil available mineral nitrogen
no.	Number
NO_3^-	Nitrate
no3_demand	Demand for nitrate
no3_tot	Nitrate available to plants
n_supply_soil	Nitrogen supply
NTRM-MSC	Nitrogen Tillage – Residue Management Model – Multiple Species Competition
n_uptake	Nitrogen uptake

Abbreviations and acronyms

obs.	Observed
P	Phosphorus
p./pp.	Page/pages
P1	Growing degree days from emergence to end of juvenile phase
P1D	Sensitivity to photoperiod
P1V	Sensitivity to vernalization
P2	Photoperiod sensitivity
P_2O_5	Phosphorus pentoxide
P5	Cumulative growing degree days from silking to maturity
P5	Grain filling duration
PAR	Photosynthetically active radiation
PHINT	Phylochron interval
PR	People's Republic
R^2	Coefficient of determination
REI	Relative efficiency index
REML	Restricted maximum likelihood
RLO	Relative land output
RMSE	Root mean square error
RNT	Relative nitrogen yield total
root_depth	Root depth
RYT	Relative yield ratio
s	Second
S	Shading
senescedwt	Senesced dry matter
sim.	Simulated
Sph	Spherical model
SPRI	Shandong Peanut Research Institute
SRAD	Daily solar radiation
SSSA	Soil Science Society of America
Stage	Growing stage
STICS	Simulateur mulTIdisciplinaire pur les Cultures Standard
SUCROS	Simple and Universal Crop growth Simulator
sw_deficit(1, 2)	Soil water deficit in different layers
sw_demand	Soil water demand

sw_supply	Soil water supply
t	Ton
TAB	Büro für Technikfolgen-Abschätzung beim Deutschen Bundestag
Tab.	Table
TDR	Time domain reflectometry
TKW	Thousand kernel weight
UN	United Nations
UNEP	United Nations Environment Programme
US/USA	United States of America
VDLUFA	Verband Deutscher Landwirtschaftlicher Untersuchungs- und Forschungsanstalten
Vs	Version
vs.	Versus
w	West plot border
Zn	Zinc
α	Probability
t	Spatial trend effect
y	Yield
μ	Intercept (column effect)
β	Fixed effect for the position
ε	Error/nugget
σ^2	Variance
°C	Degree centigrade
°E	Degree east/longitude
°N	Degree north/latitude
3D	Three-dimensional space
%	Percent
vol %	Volume percent

List of figures

		Page
Figure 1:	Development of the ratio of land per capita in China throughout the centuries (EarthTrends, 2003; Netting, 1993)	2
Figure 2:	Chinese agricultural landscape is extremely fragmented leading to the term "noodle fields" (Netting, 1993) (picture: T. Feike, 2008)	5
Figure 3:	IRTG study region 'North China Plain' (www.cossa.csiro.au/aciar/book/Overview.html)	7
Figure 4:	Field layout of the field trial in China for the years 2008 to 2009	14
Figure 5:	Field layout of the field trial in Germany for the years 2007 to 2009	16
Figure 6:	Fragmentary rear elevational view of a portion of an apparatus for sowing a second crop in a standing crop (left) (Younger, 1978). Strip intercropping appearance and practice in the USA (right) (http://www.thisland.illinois.edu/60ways/60ways_17.html)	171

List of tables

Table 1:	Geographical situation, average air temperature, average precipitation and soils of the two experimental locations in China and Germany	13
Table 2:	Description of data and parameters collected and evaluated during the experiments 2007 – 2009 in Germany	18

1 Introduction

Modern agriculture has to work out the balancing act between economy and ecology. On one hand, there is a growing demand for food supply, food access and food quality. On the other hand, there are constrains on environmental protection issues and income certitudes for farmers within a global market. High yield and sustainability are the catchphrases of production in the 21st century – and agronomy research has to provide solutions in increasingly briefer terms. In fast developing countries like China, those issues have become even more severe. Agriculture production rates and tremendously increased production levels aiming to feed the nation coming along with severe environmental pollution need adopted cropping systems for Chinese production circumstances. Thus, improved cropping strategies, which are based upon common, widespread and traditional cropping systems like intercropping, need to be developed. Taking modeling and simulation tools into account could help both to accelerate research attainments and to give a better understanding of cropping systems.

Within the last decades the Chinese agricultural development has been accompanied by serious **environmental degradation problems** (Wen et al., 1992). Exaggerated verbalized, agriculture nowadays is an attack on the environment, attacking anything that could be cut down, plowed up, pumped over, dug out, shot dead, or carried away (Hinton, 1990, p. 21). Fulfilling the premise of high productivity and food self-sufficiency within a country where the average population density is four times higher than global population density (Prabhakar, 2007) and, in contrast, the land per capita is only 0.11 ha (EarthTrends, 2003) (Figure 1) resource use efficiency takes on a dimension beyond sustainability. That is what Evans (1998) called the sustainability dilemma when agriculturists are often torn between their concerns about the need for greater food production and the need to conserve what is left of nature and of the resources of agriculture for future generations.

1. Introduction

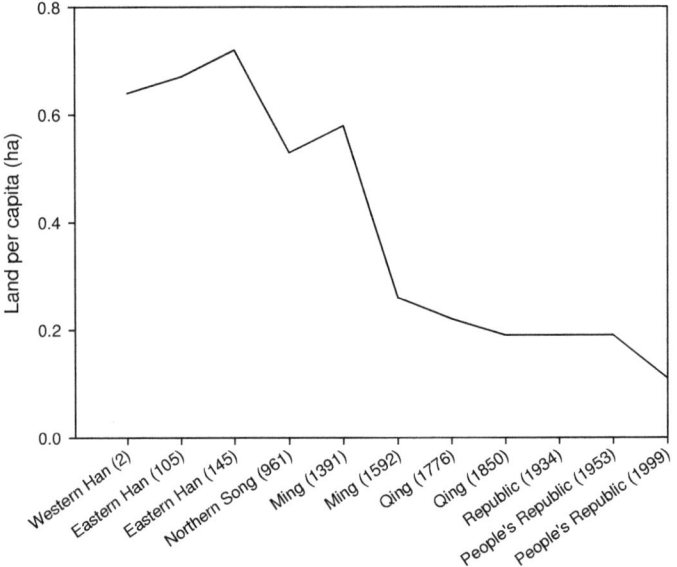

Figure 1: Development of the ratio of land per capita in China throughout the centuries (EarthTrends, 2003; Netting, 1993)

THE PARADIGM OF SUSTAINABILITY

Looking upon soil erosion, degradation and desertification in China as an example, five billion tons of soils are eroded each year, resulting in a loss of nutrients associated with organic matter equal to twice the national production of chemical fertilizers (Tong et al., 2003). 1/6 of the total land area and 1/3 of arable land in China are presumed to be eroded; the total soil loss in one year was 20 % of the total soil loss in the world, and deserts are expected to expand at a rate of 1560 km² per year (Wen et al., 1992; Yunlong and Smit, 1994).

In addition, China ranks first in the world with respect to the amount of nutrient fertilizers used. In the North China Plain, the N fertilizer application rate for a maize/wheat double cropping system is 400-600 kg N ha^{-1} per year (Fang et al., 2006). For peanut (legume-) production, recommended fertilizer amounts range between 3.75-7.5 t farmyard manure combined with additional 60-100 kg N (Liang, 1996; Cai et al., 1996) or, for a peanut/wheat relay intercropping, 60 m³ farmyard manure combined with 300-375 kg urea or 15-22.5 t farmyard manure combined with 150 kg urea provided as second application in spring (Zhang et al., 1996). As a result, nitrate leaching is a big issue in

China. Fang et al. (2006) reported that 35 % of total N application in 1987 was lost due to leaching. A survey in northern China indicated that the maximum nitrate content in ground and drinking water of 109 mg l^{-1} exceeded accepted health standards by an average of 50 mg l^{-1} (Wang et al., 2009; Zhang et al., 1996).

Taking these two aspects – soil erosion and N use efficiency – into consideration, it is obvious how important it is to develop technically feasible agricultural systems, which are friendly towards the natural environment and which are predisposed towards farmers' adoption and use (Sanders, 2000). The question is how to combine the welcome reduction in hunger and malnutrition, how to earn cash and how to reduce the threats to ecological sustainability over time? Prabhakar (2007, p.18) pointed out that revising the existing cropping patterns and systems is needed, and as monocropping means higher risk, in terms of income security, nutritional diversity in rural areas, and possibility of severe impacts to large areas due to pest and disease outbreak in changing climate, mixed and intercropping practices are the only alternative that has multiple benefits.

In the **intercropping** research context the following benefits appear repeatedly (Knörzer et al., 2009): maximized land use, several harvests per year, yield stability, increased resource use efficiency, reduced soil erosion and leaching, and reduced pests and diseases. As a scale unit for maximized land use, the land equivalent ratio (LER) is most commonly used. It defines the ratio of land for monocropping necessary in comparison to intercropping. In his appendix, Innis (1997) gave a compilation of the results of several hundred experiments, which have been done on intercropping using the LER. In most of those studies, maize or legumes or even both were involved. In some of them, the LER of monocrops, which is 1, was almost doubled for intercrops, showing that intercropping had higher yield expectancies than its monocropping equivalents for the same proportion of land. Not only yield in total might be higher in intercrops, but yield stability also, as one crop failed because of climatic or pest and disease reasons, there was still another crop left. In addition, relay intercropping, where the maturing annual plant is interplanted with seeds of the following crop, allows an additional harvest within the growing season as the overlapping growing period elongates the growing period of the second crop. In the Northern and North-Eastern provinces of China, relay intercropping systems with wheat/maize (Li, 2001; Li et al., 2001; Yamazaki et al., 2005; Wang et al., 2009), wheat/peanut (Cai et al., 1996; Liang, 1996; Zhang et al., 1996) and wheat/cotton (Li, 2001; Zhang et al., 2008) are widely and successful practiced, indicating that intercropping is not restricted to cereal/legume combinations or maize as a major component. Indeed, the species spectrum within intercropping systems in China is extensive and includes wheat, maize, cotton, green manure, soybean, sweet potato, rape, peanut, broomcorn millet, bean, buck wheat, millet, tobacco, sorghum, rice, cassava, garlic and a great variety of vegetables (Meng et al., 2006; Li, 2001).

Literature gives evidence that the advantage of intercropping compared to monocropping is the increased nutrient and light use efficiency (Fukai and Trenbath, 1993; Inal et al., 2007; Li et al., 2003; Song et al., 2007; Vandermeer, 1989; Zhang et al., 2004). Competition and facilitation could be balanced by choosing appropriate components and a mutually beneficial timing. Permanent covered soil is less susceptible to erosion, and as example from the middle reaches of Heihe River in the Hexi Corridor region showed, a switch to intercropping intimated a reduction in soil wind erosion and a halt in sand entrainment (Su et al., 2004). In this region it occurs that dust transport from farmlands is about 4.8 to 6.0 million tons per year and consequently higher than that of sandy desert dust transport in the same region. In addition, several studies indicated that intercropping reduced nitrate leaching (Li et al., 2005; Song et al., 2007; Ye et al., 2005).

For Chinese farmers the reasons for intercropping are more driven by basic principles than by sophisticated aspects like sustainability or protection of environment. On the one hand there is a long tradition of those systems being passed on from generation to generation, and on the other hand there is China's situation of land shortage and abundant labor – and intercropping systems in China are extremely labor intensive. Average farm size is 0.1-0.5 ha with fields being extremely small. Farmers have to make the most of their small plots in space and time which are carefully economized in the Chinese Intensive System (Netting, 1993). First priority is maximizing land use and income security. Agricultural landscape is fragmented leading to the term "noodle fields" (Netting, 1993) (Figure 2). This **smallholder farming** in combination with labor and time intensive cropping systems are stigmatized to be old-fashioned and less efficient than those agro-industries in the United States of America or Australia.

On a political level, Netting (1993, p. 21), who studied smallholder systems very briefly, stated for both, the socialists and the communists on the left and the free-market capitalists on the right, the agreed-upon path to agricultural development has been the large-scale, mechanized, energy-dependent, scientific, industrialized form. In contrast, tradition or cultural values are said to be responsible for economic irrationality. Accordingly, is intercropping in its existing form a dwindling system? The paradigms of sustainability and biodiversity convey not only the resource protection consciousness of industrialized countries' society, but also their forms of agricultural practice. Netting (1993, p.151) argued that the productivity of land declined sharply on large farms and the risks of deforestation, erosion, and possibly permanent environmental degradation increased. In the past, only the smallholder intensification increased output per unit of land, while conserving natural resources. Nowadays, China's smallholders face rapidly developing agricultural production with high chemical fertilizer and pesticides input, mechanization and monocropping of cash crops. The combination of scarce land, cash and technological progress shows an ever increasing trend to deforestation, erosion, and potentially permanent environmental degradation. Across China's long agricultural history it has always been assumed that farming activities should be in accordance with seasons, climate, soil conditions, and nutrient input. And it has been in that long agricultural history that intercropping developed and always played an important role. This does not mean that Chinese agriculture has to go back to its roots, but it shows the potential and the importance of typical Chinese cropping systems to be improved, adapted, and refined.

Figure 2: Chinese agricultural landscape is extremely fragmented leading to the term "noodle fields" (Netting, 1993) (picture: T. Feike, 2008)

ECOLOGICAL AWARENESS AND RESEARCH

Within the last two decades, there is an increasing awareness of environmental pollution and degradation due to agricultural production in China. Studies on agricultural development and land use in China mostly open with the problem of China's population density and its lack in arable land: Chinese farmers have to feed 22 % of the world's population with only 7 % of the world's arable land (Li, 2001; Liu et al., 2001; Tong et al., 2003; Wen et al., 1992). Those studies outline the serious challenges China will face on its still long way to truly sustainable agriculture having enormous environmental costs to carry on its shoulder. Thus, agricultural research aims to cope with those constraints leading to a vast variety of national and international projects such as the International Research Training Group (IRTG) "Sustainable Resource Use in the North China Plain", within which this research was based.

The project is entitled **"Modeling Material Flows and Production Systems for Sustainable Resource Use in Intensified Crop Production in the North China Plain"** and was established by the Deutsche Forschungsgemeinschaft (DFG) and the Chinese Ministry of Education at the Universität Hohenheim (Stuttgart) and the China Agricultural University (Beijing) in June 2004. It will run until 2013. With a background of high level production intensities, raising productivity, increasing food demand of a growing population with increasing living standards and coincidently a lack of sustainability the IRTG's aims were to develop environmentally, economically and socially sustainable cropping and management systems. This requires clearly identifying, measuring, and modeling the related material flow effects in cropping systems on field, farm, and regional levels (https://rtgchina.uni-hohenheim.de/start.html?&L=1). Eleven subprojects from various disciplines (soil science, plant nutrition, plant ecology, plant production, plant breeding, weed science, agricultural engineering, farm management, agricultural informatics and rural policy) are involved and several dissertations on those fields have been released so far.

Research has been focused on the North China Plain (NCP), also known as the Huang-Huai-Hai Plain (Figure 3). It is the so called granary of China, where approximately one-fourth of the country's total grain yield is produced, with wheat, maize and cotton as major crops. It is located in the country's North-East (32°-40°N, 100°-120°E) and composed of the piedmont diluvial-alluvial plains of Taihang Mountains and Yanshan Mountains and the alluvial plains of the Huanghe and Haihe River (Liu et al., 2001; Xi, 1989). The climate is characterized as a continental monsoon climate with an average temperature ranging between 10-14°C and an average precipitation between 500-800 mm (Liu et al., 2001). The NCP is a major base of agricultural production in China and plays a vital and strategic role in the development of the national economy (Xi, 1989, p. 16). Therefore, the region could be ranked as a hot spot for pollution caused by agricultural production as the problems herein reach a certain point of culmination.

1. Introduction

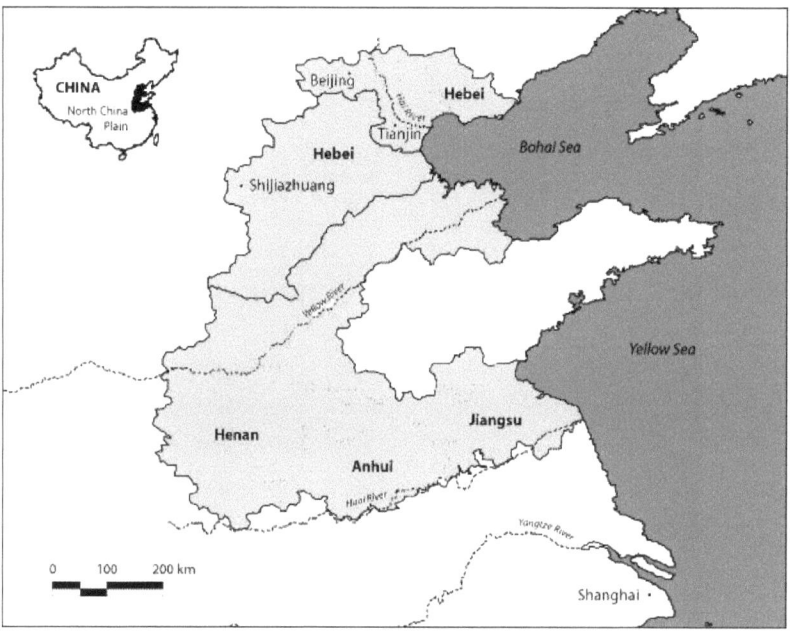

Figure 3: IRTG study region 'North China Plain' (www.cossa.csiro.au/aciar/book/Overview.html)

Within the IRTG project, the subproject 'crop production' was included. Based on a process-oriented modeling approach, the subproject emphasis' was on the evaluation of cropping system prototypes with special regard to intercropping (Project Proposal, 2007). Several crops were involved: spring maize (*Zea mays*), winter wheat (*Triticum aestivum*), peanut (*Arachis hypogaea*), and pea (*Pisum sativum*). The aim was to explore the potential and the possibilities of intercropping and the competition and facilitation effects of those systems in order to create new methodologies for a better understanding of and improvement of existing cropping systems. Necessary methods and basic approaches for the description of the relevant indicator parameters were designed and transferred into a thorough modeling approach (Project Proposal, 2007) for the simulation of intercropping experiments. Datasets collected from both, German and Chinese field experiments 2007-2009, were used to develop a shading algorithm to be incorporated into the DSSAT (Decision Support System for Agrotechnology Transfer) crop growth model (Jones et al., 2003). So far, the current version of DSSAT aims to simulate monocrop production systems (Jones et al., 2003) and does not take interspecific competition into account. Intercropping could not be modeled or simulated.

1. Introduction

DECISION SUPPORT SYSTEM

> "This decision support system is designed to answer "what if" questions frequently asked by policy makers and farmers concerned with sustaining an economically sound and environmentally safe agriculture. Sustainable agriculture requires tools that enable decision makers to explore the future. A decision support system must help users make choices today that result in desired outcomes, not only next year, but 10, 25, and 59 or more years into the future (Tsuji et al., 1994)".

In their introduction for the 3rd DSSAT volume manual, Tsuji et al. (1994) clearly pointed out, that nowadays agriculture and agriculture research are strongly related to sustainability and consequential relevant decisions for the future. Since the outcomes of crop growth models in the late 1960s and early 1970s (Acock, 1989; Jones, 1998), research questions and topics changed from "what is the explanation for what has been" to "how can the knowledge be applied for predicting future scenarios". **Computer simulation models** of the soil-crop-atmosphere system can make a valuable contribution to both furthering the understanding of the processes determining crop responses and predicting crop performance in different areas and different cropping systems (Project Proposal, 2007). Understanding the underlying processes of plant growth within a given environment and a given management system can be used for prediction of plant growth in different environmental conditions, even with respect to climate change, the introduction of new species into a cropping system, and different management strategies. Understanding and prediction might be the push back factors for managing and controlling and hence, offering the opportunity for change and improvement.

There are various types of models: statistical and kinetic, static and dynamic, empiric and mechanistic, deterministic and stochastic as well as one-dimensional and multi-dimensional. For modeling and simulating cropping systems, process-oriented crops growth models are preferred, such as the DSSAT model, which was used predominantly for this study, as well as the APSIM model (Agricultural Production Systems Simulator) (McCown et al., 1995, 1996).

Process-oriented means a programming paradigm that separates the concerns, shares logically data structures and the concurrent processes that act upon them (Ericsson-Zenith, 1992). As a result, process-oriented models are suitable for large scale applications that partially share common data sets. In agricultural practice, both crop growth and applied management strategies such as rotations, tillage, fertilizers, and plant protection are important elements of success. Thus, a systems approach is needed, where a system is defined as a part of reality which can be delimited from its surroundings, for example soils, plant canopies, and atmosphere. Within those environmental and biophysical plant-soil-atmosphere models, various single heat- and mass-transfer mathematical models are combined to map the exchange processes between organisms and their surroundings (Campbell and Norman, 1998).

The **DSSAT** Vs 4.5 is able to simulate crop growth development of 30 different species for uniform areas of land under prescribed or simulated management as well as dry matter- and grain yield. It can also simulate changes in soil water, carbon and nitrogen stocks and fluxes (Jones et al., 2003). It is organized in different and independent acting modules which operate together. In its center the crop simulation models with underlying genotype characteristics are placed surrounded by weather and soil models as well as experimental conditions and measurements options each with individual databases defining and describing basic conditions, dependencies and functions.

Jones et al. (2003) provided a comprehensive list of modeling and simulation studies wherein the DSSAT model has proven to be a robust model to simulate monocropping situations and experiments in uniform, spatial as well as modified environments all over the world. In addition, there are various models dealing with simulating intercropping scenarios, as shown in chapter two within this dissertation (Knörzer et al., 2010).

Some models for interspecific competition only deal with a few aspects of intercropping, e.g. mortality of individual plants in a stand (Yokozawa and Hara, 1992) or radiation transmission (Tsubo and Walker, 2002; Tsubo et al., 2005), and not with the cropping system itself. In addition, modeling intercropping is often based upon the assumption that the competing plants have a common pool (soil water and nutrients and solar radiation) for their supply of growing factors, ignoring that most intercropping systems are not species mixtures within a given area, but separated and alternating strips of different crops. Leaf area index and plant height are the driving forces for simulated plant growth in intercropping model approaches with e.g. the ALMANAC (Kiniry and Williams, 1995) or APSIM (Carberry et al., 1996; Nelson et al., 1998) model. These approaches are limited with respect to intercropping systems where both species are planted at the same time and in an appropriate plant density that allows the understorey plant to gain sufficient light for growth. In relay intercropping systems, where one plant is far ahead in its development before the companion plant is sown, these models are of limited use.

Thus, a new approach is presented in chapter two and five as part of the topic design, model and evaluation of improved cropping strategies in intercropping systems. Therefore, the DSSAT model was chosen as the model capable of simulating various different crops, having proven to be robust in agriculture systems modeling, offering the possibility to test a new competition algorithm, and to broaden the view on intercropping in a spatial dimension.

1. Introduction

OUTLINE AND HYPOTHESES

Ecology meets agronomy in patchy agricultural landscapes (Knörzer et al., 2010). Intercropping has not been analyzed from the point of view that fragmented agricultural landscapes with predominant smallholder farming correspond to intercropping at a larger scale, defining the sum of small fields next to each other as intercropping. Hence, the theory of borders or boundaries from ecology can be transferred to agronomy: The border and the two edges appear as a consequence of the interaction and constitute what is called a boundary (Fernández et al., 2002). Within each small field there is a yield distribution between border rows and centered rows comparable to intercropping and monocropping as intercropping depends upon field border effects. Thus, modeling intercropping could be or interpreted as modeling field boundaries.

To analyze, design, evaluate and in the end model intercropping within an agricultural landscape where smallholders are predominant, the following **elementary questions** have to be answered:

- What are the status quo and the potential of intercropping in China?
- Are common mixed models applicable for intercropping?
- Do we have to broaden our view on intercropping to a spatial dimension?
- How is competition for solar radiation modeled, and is it the driving force within intercropping systems?
- How shall interspecific competition effects be weighted?
- Can intercropping be modeled with DSSAT using simple algorithms?

To answer these questions the underlying **hypotheses** evoked: Crop growth models can be used to simulate intercropping systems by implementing site-specific modeling and by evaluating general competition algorithms. Phenological, morphological and physiological differences between different species handled as intercropping partners will increase beneficial and synergistic effects concerning yield and resource use efficiency.

Covering the full range of intercropping from intercropping distribution and intercropping practice in China (macro level) to site-specific simulation of a single intercropping system (micro-level), the dissertation is structured according a stepwise approximation from the macro to the micro level.

In the **first chapter**, a literature review about the status quo of intercropping of cereals in China was done. It was a first approach to identify and classify different agro-climatic regions in China according to their intercropping systems, their distribution, their frequency of occurrence, their species combination, and their degree of intensification. In addition, the overview over the long and successful history of intercropping combined with the actual research effort in Chinese

1. Introduction

intercropping systems led to an assessment of intercropping sustainability and its future potential within Chinese agriculture.

In the **second chapter**, crop growth models were reviewed which are already able to simulate interspecific competition or intercropping, or model approaches which have introduced basic and elementary competition algorithms. In that context, around 20 different models were presented. Most of these models simulated intercropping by using either the turbid layer medium analogy or the principle of reducing leaf area index of the understorey species as far as the dominant species reached a special height. As the CERES models within DSSAT do not model plant height a different approach for simulating changed solar radiation within intercropping systems has to be developed. Thus, a unique and contradicted model approach was used by evaluating a shading algorithm taking the neighboring plant height and its proportional shading potential to the target plant into account.

Chapter three deals with the problem how to analyze intercropping systems adequately. As intercropping is mostly practiced as row or strip intercropping – on the fields as well as in intercropping experiments –, there is a problem in statistical data analysis: Strip intercropping systems lack in a basic analytical principle, namely randomization. Thus, usually applied analysis of variance (ANOVA) with a baseline model might estimate significant differences too optimistically. Spatial models have to be applied helping to improve the model fit and to analyze intercropping as well as field boundary effects.

As China is the largest peanut producer in the world, and peanuts were included into the Chinese field trials, **chapter four** was added as some kind of excursus into the dissertation. Four major peanut producing provinces were chosen for a modeling study in order to simulate large area yield estimation and to evaluate potential yield with respect to average rainfall. Anhui, Henan, Hebei and Shandong are located in the North-Eastern part of China. In these regions, drought stress between germination and pod setting could be severe and yield declines because of uneven rainfall and climate variability.

In **chapter five**, the intercropping modeling approach with the DSSAT model, first presented in chapter two, was further tested. In more detail, the shading algorithm for the wheat and maize crops grown as inter- and monocrops was described and introduced into the DSSAT model as modified weather input. In addition, microclimate influence, such as solar radiation, wind speed and soil temperature, and its distribution over the growing season in intercropping in comparison to monocropping systems were studied more closely. Nitrogen dynamics based upon soil temperature differences as well as CO_2 dynamics based upon windbreak effects of a taller neighboring plant might often be small or subtle, but their spatial variability in conjunction with the heterogeneity of plant canopies can be considerable and constitute the starting point for modeling intercropping.

In **chapter six** there is a switch from the DDSAT to the APSIM crop growth model. Both models are process-oriented plant-soil-atmosphere models and operate similarly. Previous versions of APSIM were based upon the cereal models within DSSAT and thus, both models calculate crop phenology based upon thermal time unit requirements. In addition, the models have some bio-physical algorithms in common. Nevertheless, the DSSAT intercropping approach presented in this thesis differs substantially from that in APSIM (Carberry et al., 1996). Incoming solar radiation and photosynthetic active radiation (PAR) are modeled in different ways, especially with respect to competition for solar radiation in intercrops. The comparison between DSSAT and APSIM concerning their intercropping modeling approach was not intended to be especially related to accuracy or to test whether the one or the other has been more suitable for modeling intercropping. The comparison was done order to increase the understanding of modeling competition and to review the general modeling of competition for solar radiation and hence, further research emphases.

Intercropping is a good deal more than the survival of the fittest or a blind alley for overcoming modern high-tech agriculture. Each intercropping system is a system on its own, sensitive balanced between competition and facilitation out in the fields and between benefits and limitations for the farmers' practice. Thus, generating the basic methodology for using a process-based simulation model for the design and strategic planning of intercropping systems under different agro-climatic conditions could contribute to a further understanding and improving of intercropping systems and, hence, to intercropping winning its due acceptance.

2 Field Experiments conducted for publications

To study intercropping effects in comparison to monocropping and to develop a suitable dataset for modeling intercropping in DSSAT, field experiments were conducted in China (experiment I) as well as in Germany (experiment II) during the years 2007 and 2009 (Table 1). The experimental design used at both locations was similar in order to apply similar statistical methods for analyzing and to study similar processes involved, but the intercropping components differed. In China, a maize/peanut intercropping system, especially widespread in northern parts of China, and in Germany, a maize/pea as well as a maize/wheat intercropping system was used.

To evaluate a basic dataset that can be used for introducing interspecific competition into the model, it was important to design a simple experiment allowing effects occurring in intercropping and monocropping during the growing season to be assigned to the relative cropping system. Thus, an experimental layout without different conventional treatments, e.g. fertilizer amount within plots, was chosen and arranged as a non-randomized complete block design with four replications. Randomization was not possible as intercropping experiments need alternating plots, strips or rows. In contrast, the distance to plot border where intraspecific competition turns more and more into interspecific competition was taken as treatment. As a result, one or each experimental plot within a block contained one species and was divided into several subplots or strips, each defined by their distance to the plot border and each containing several crop rows. Data was collected within subplots.

Table 1: Geographical situation, average air temperature, average precipitation and soils of the two experimental locations in China and Germany

	Wuqiao (China)	Ihinger Hof (Germany)
Geographical situation	37.3°N and 116.3°E	48.46°N and 8.56°
Average air temperature	13.1°C	7.9°C
Average precipitation	562 mm	690 mm
Soils	Alluvial soils sandy clay	Keuper with loess layer silty clay loam

EXPERIMENT I: FIELD TRIALS IN CHINA 2008 TO 2009

Experiment 1 was conducted in northeast China (37.3°N and 116.3°E) at the China Agricultural University experimental station in Wuqiao, Hebei province, during the years 2008 and 2009. In Wuqiao, the average rainfall per year is 562 mm; the average temperature per year is 13.1°C. In this region, mainly alluvial soils occur with a disposition to salinization.

The experiment was designed in four non-randomized complete blocks with alternating plots of spring maize and peanut in each block (Figure 4). Maize and peanut were sown in May 2008 and 2009 with a row spacing of 60 cm and 30 cm, respectively, and a sowing depth of 5 cm. Plant density of spring maize was seven plants m^{-2}, plant density of peanut was 30 plants m^{-2}. No N fertilizer was applied. Pesticides were given once a week after emergency of maize until stem elongation. Spring maize and peanut were harvested between mid and end of September 2008/09. Both crops were sown in alternating plots 8 m wide and 20 m long. Each plot was divided into several subplots in dependency from the distance to the plot border. The plots were big enough to reflect monocropping within the central subplot. In the Chinese experiment, every row equaled a subplot. Thus, data collection was done row by row via square meter-cuts in order to detect differences within the positions of the plots.

During the growing season, three 0.6 m^2 cuts at different maize and peanut growth stages were taken in each subplot to determine dry matter accumulation. In addition LAI as well as plant height were measured over the growing season. At maturity, 1.2 m^2 cuts were done to determine final dry matter and grain yield of both crops. In addition, the number of kernels per ear and the thousand kernel weight (TKW) were measured for maize.

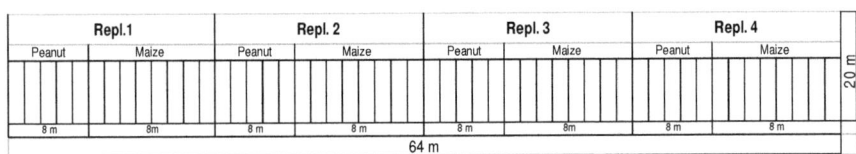

Figure 4: Field layout of the field trial in China for the years 2008 to 2009

EXPERIMENT II: FIELD TRIALS IN GERMANY 2007 TO 2009

The second and similar experiment was conducted in southwest Germany, at the Universität Hohenheim experimental station 'Ihinger Hof' during the years 2007 to 2009. The station is located 48.46°N and 8.56°E and has an average temperature of 7.9°C per year and an average rainfall of about 690 mm per year. Dominant soils are keuper with loess layers.

The experiment comprised a maize/wheat intercropping system as well as a maize/fieldpea intercropping system, each within a not-randomized complete block design and four replications (Figure 5). Thus, each replication contained complete blocks of both systems and each system's replication consisted of four plots. Two plots were used for time harvests and weekly measurements during the growing season and another two plots were used for the final harvest. The two species were planted in an alternate pattern. Each plot was 10 x 10 m² for wheat as well as for pea, and 12 x 10 m² for maize, respectively, and included five subplots (5 x 2 x 10 m²) for wheat and pea, and eight subplots (8 x 1.5 x 10 m²) for maize. Within those subplots, data was collected (Table 2) in order to detect crop performance differences between different distances from the plot border. The plots were big enough to reflect monocropping within the central subplot. Row orientation was from north to south.

In both years, the previous crop was sugar beet and soil preparation and sowing was done by a reduced tillage system. Plant protection was carried out according to 'Good Agricultural Practice'.

The wheat variety 'Cubus' was sown in October 2007 and 2008 with a row spacing of 13 cm and a plant density of 300 plants per m². Maize ('Companero') was sown in May 2008 and 2009 with a row spacing of 75 cm and a plant density of 10 plants per m². The fieldpea variety 'Hardy' was sown between end of March and beginning of May 2008 and 2009 with a row spacing of 13 cm and a plant density of 70 plants per m². Harvest of wheat and pea was at the end of July. Maize was harvested in October.

Wheat was fertilized with 160 kg N ha^{-1}, split into three dispensations (60/60/40) of Nitro-chalk. Maize was fertilized once with 160 kg N ha^{-1} (ENTEC). Fieldpea as a leguminous plant was not fertilized at all.

2. Field Experiments conducted for publications

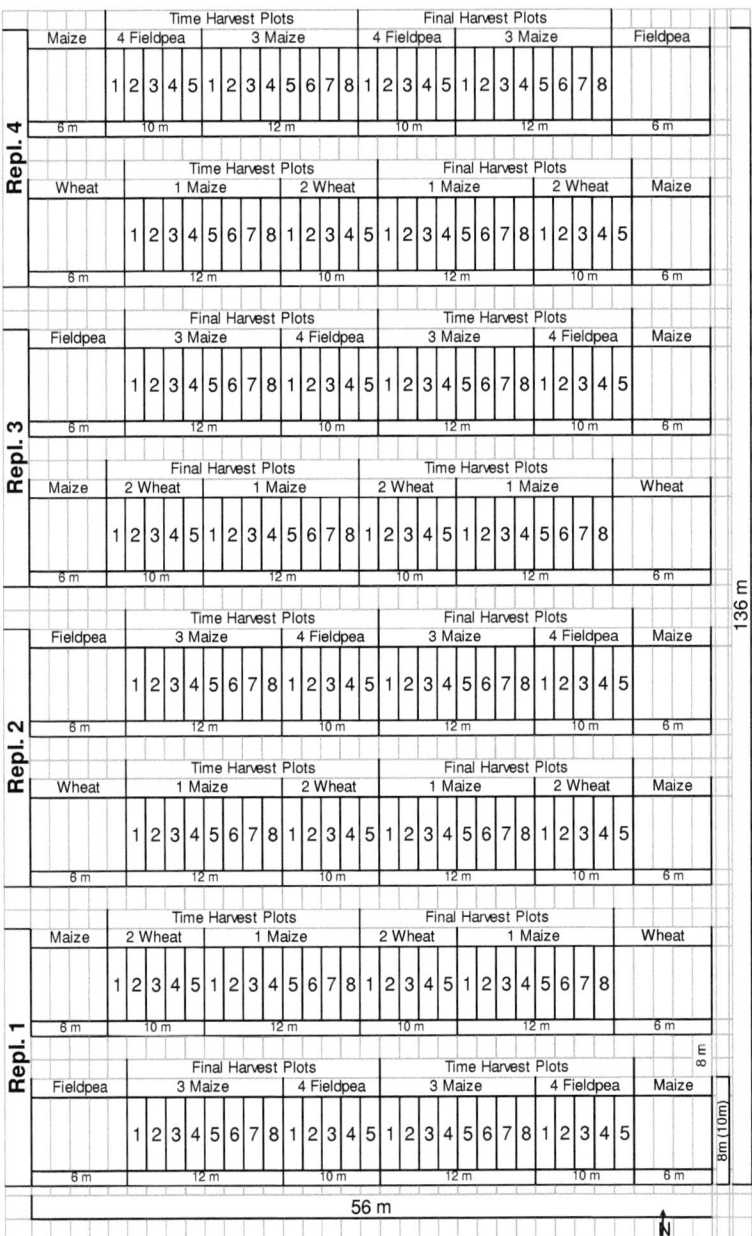

Figure 5: Field layout of the field trial in Germany for the years 2007 to 2009

2. Field Experiments conducted for publications

During both growing seasons, neither water nor nitrogen (N) stress occurred so that differences in plant growth and yield performance could be attributed onto intra- and interspecific competition. Weekly measurements of soil water content between end of May and beginning of July with a Trime-TDR-system (time domain reflectometry) from IMKO GmbH (Ettlingen/Germany) indicated that no water shortage occurred during the two growing seasons.

Three temporal harvests according to the DSSAT guide were carried out as square meter cuts, and grain yield, dry matter, and nitrogen concentration of plants were analyzed 2007 - 2009. Nitrogen concentration was determined with the NIRSystems 5000 (ISI-Software, USA). In addition, N_{min} content of soil was determined at the beginning of the growing season and after harvest according to VDLUFA methods (VDLUFA, 1991). In addition, in the year 2009, five N_{min} samples were taken weekly between May and June in wheat, three in pea and two in maize, in order to detect differences between N budgets within different subplots. Soil temperature was determined in 2 cm soil depth with the testo 925 (Testo AG, Lenzkirch/Germany) on a weekly basis as well as solar radiation with the AccuPAR LP-80 (UMS, München/Germany) and the testo 545 (Testo AG, Lenzkirch/Germany) 2007 - 2009. Wind speed was measured in the year 2009 above the canopy with anemometer compact (Thies Clima, Göttingen/Germany). Growing stages according to the BBCH scale (Meier, 1997) and plant height were reported on a weekly basis. After the final harvest, yield and yield components like thousand kernel weight (TKW), tiller number, ears or pods per plant as well as N concentration and N uptake were determined for all crops.

2. Field Experiments conducted for publications

Table 2: Description of data and parameters collected and evaluated during the experiments 2007 – 2009 in Germany

Data / Parameter	Determination
Plant growth stage	2007 -2009: documented weekly from emergence to maturity using the BBCH scale (Meier, 1997)
Plant density	2007 – 2009: wheat/pea: 0.7 m² counted (2 rows x 2.75 m x .0,13 m) maize: 1 m² counted (2 rows x 0.66 m)
Plant height	2008 – 2009: measured weekly between May and July
N_{min}	2007 – 2009: determined at the beginning of the growing season and after harvest according the VDLUFA methods (VDLUFA, 1991) 2009: between May and June weekly samples (5 x wheat, 3 x pea, 2 x maize)
Soil moisture	2008 – 2009: between May and July weekly measurements with the Trime-TDR-system (IMKO GmbH, Ettlingen/Germany)
Solar radiation	2008 – 2009: measured weekly between April and July with the AccuPAR LP-80 (UMS, München/Germany) and the testo 545 (Testo AG, Lenzkirch/Germany)
Soil temperature	2008 – 2009: measured weekly between May and June with the testo 925 (Testo AG, Lenzkirch/Germany)
Wind speed	2009: measured above the canopy continuously during the growing season with anemometer compact (Thies Clima, Göttingen/Germany)
LAI	2008: between May and June weekly destructive LAI determination within wheat with the LI-3100 Area Meter (LI-COR, Lincoln/Nebraska USA) 2008 – 2009: between May and July weekly non-destructive LAI determination with the LAI-2000 Plant Canopy Analyzer (LI-COR, Lincoln/Nebraska USA)
N/C concentration	2008 – 2009: time and final plant harvest samples were analyzed with the NIRSystems 5000 (Foss, Rellingen/Germany; ISI-Software/USA)

2. Field Experiments conducted for publications

Data / Parameter	Determination
Time harvests - dry matter -	2008 – 2009: three square meter cuts during the growing season maize: 1st at BBCH 16, 2nd at BBCH 63, 3rd at BBCH 85; wheat: 1st at BBCH 13, 2nd at BBCH 65, 3rd at BBCH 85; pea: 1st at BBCH 14, 2nd at BBCH 67, 3rd at BBCH 79
Time harvests - grain yield -	2008 – 2009: three samplings every second week after anthesis wheat: 10 ears/subplot cut, dried and weighted maize: 4 ears/subplot cut, dried and weighted
Final harvests	2008 – 2009: square meter cuts for determining dry matter; harvesting of subplots with Hege 180 for determining grain yield (15.5 m² of wheat and pea, 11 m² of maize)
Yield components	2008 – 2009: TKW ears/pods per plant kernels per pod number of tillers per m²

3 Publications

The thesis consists out of six chapters which have been published as papers to peer-reviewed and international high standard referenced journals or books. For citation of chapters I-VI, please use the references given below.

CHAPTER I:
Knörzer, H., Graeff-Hönninger, S., Guo, B., Wang, P., and Claupein, W. (2009): The rediscovery of intercropping in China: a traditional cropping system for future Chinese agriculture. Springer Series: Sustainable Agriculture Reviews 2: Climate Change, Intercropping, Pest Control and Beneficial Microorganisms, ed. by E. Lichtfouse. Springer Science+Business Media, Berlin, pp. 13-44.

CHAPTER II:
Knörzer, H., Graeff-Hönninger, S., Müller, B.U., Piepho, H.-P., and Claupein, W. (2010): A modeling approach to simulate effects of intercropping and interspecific competition in arable crops. International Journal of Information Systems and Social Change 1 (4), pp. 44-65.

CHAPTER III:
Knörzer, H., Müller, B.U., Guo, B., Graeff-Hönninger, S., Piepho, H.-P., Wang, P., and Claupein, W. (2010): Extension and evaluation of intercropping field trials using spatial models. Agronomy Journal 102 (3), pp. 1023-1031.

CHAPTER IV:
Knörzer, H., Graeff-Hönninger, S., and Claupein, W. (2010): Model-based approach to quantify and regionalize peanut production in the major peanut production provinces in the People's Republic of China. GI-Edition - Lecture Notes in Informatics „Precision Agriculture Reloaded – Informationsgestützte Landwirtschaft", pp. 101-104.

CHAPTER V:
Knörzer, H., Grözinger, H., Graeff-Hönninger, S., Hartung, K., Piepho, H.-P., and Claupein, W. (2011): Integrating a simple intercropping algorithm into CERES-wheat and CERES-maize with particular regard to a changing microclimate within a relay-intercropping system. Field Crops Research 121 (2), pp. 274–285.

CHAPTER VI:
Knörzer, H., Lawes, R., Robertson, M., Graeff-Hönninger, S., and Claupein, W. (2011): Evaluation and performance of the APSIM crop growth model for German winter wheat, maize and fieldpea varieties in monocropping and intercropping systems. Journal of Agricultural Science and Technology 5 (12), pre-print version.

4 Chapter I:

The rediscovery of intercropping in China: a traditional cropping system for future Chinese agriculture

PUBLICATION I:

Knörzer, H., Graeff-Hönninger, S., Guo, B., Wang, P., and Claupein, W. (2009): The rediscovery of intercropping in China: a traditional cropping system for future Chinese agriculture. Springer Series: Sustainable Agriculture Reviews 2: Climate Change, Intercropping, Pest Control and Beneficial Microorganisms, ed. by E. Lichtfouse. Springer Science+Business Media, Berlin, pp. 13-44.

The first chapter marks the status quo of intercropping practice and intercropping research in China as well as a starting point of future research within that topic. The term 'rediscovery' might be misleading as various multiple cropping systems, especially intercropping, are existing, practiced and developed in China, and the agricultural history of those systems is as long as diversified. Nevertheless, the term should acknowledge the vulnerability of a cropping system often considered as a typical cropping pattern for African and Asian smallholder farming, and stigmatized as old-fashioned and hand-labour dominated. Thus, the chapter should give a brief overview over the potential of intercropping regarding its tradition and distribution as well as its contributions for an increase in sustainability within existing cropping systems. In addition, integrating intercropping into modern scientific research methods led to several international scientific publications which increased the knowledge of species combination and interspecific competition in intercropping. Basically, the study was a literature review aiming to answer the question: 'What were and what are the contributions and extent of intercropping of cereals for Chinese agriculture in past, present and future?' The spectrum of intercropping is even broader than described here where the main emphasize was laid upon intercropping of cereals leaving aside vegetables and trees. There is obviously a vast amount of different intercropping systems in China. But in comparison to some African countries or India, the systems seem to be less documented and studied. Still, there are many black dots on the map of Chinese intercropping. The study is a first approach to comprehend basic definitions, tradition, provincial practice, and potential as well as international research efforts on intercropping of cereals in China.

4. Chapter I

The Rediscovery of Intercropping in China: A Traditional Cropping System for Future Chinese Agriculture – A Review

Heike Knörzer[1], Simone Graeff-Hönninger[1], Buqing Guo[2], Pu Wang[2], and Wilhelm Claupein[1]

[1] Univ. of Hohenheim, Institute of Crop Science, Fruwirthstrasse 23, 70593 Stuttgart, Germany
[2] China Agricultural Univ., College of Agronomy and Biotechnology, No. 2 West Yuan Ming Yuan Rd., Beijing 100094, China

Article from the Springer Series: Sustainable Agriculture Reviews 2: 13-44 (2009), with permission of Dr. Eric Lichtfouse, Editor of Sustainable Agriculture Reviews, copyright © 2009 by Springer Science+Business Media, Berlin.

ABSTRACT

Intercropping has a 1000-year old history in Chinese agriculture and is still widespread in modern Chinese agriculture. Nowadays, agricultural systems in China are stigmatized to exhaust high levels of input factors like N fertilizer or irrigation water and to contribute severely to environmental problems like desertification, river eutrophication, soil degradation and greenhouse effect. In this context, monocropping systems have to be revised and may not be the best performing systems any more, considering sustainability, income security and nutritional diversity in rural areas. Therefore, intercropping systems offer alternatives for a more sustainable agriculture with reduced input and stabilized yield. Especially in the last decade this cropping system has been rediscovered by scientific research. Studies showed increased yield of maize and wheat intercropped with legumes: chickpea facilitates P uptake by associated wheat, maize intercropped with peanut improves iron nutrition and faba bean enhances N uptake when intercropped with maize. China's intercropping area is the largest in the world. Nevertheless, there are only few international studies dealing with intercropping distribution, patterns and crops. Most studies deal with nutrient-use efficiency and availability. This study is a first approach to gain an overview of intercropping history, basic factors about interspecific facilitation and competition and distribution of Chinese intercropping systems. Finally, four intercropping regions can be distinguished and are explicitly described with their intercropping intensity, potential and conditions.

Keywords Arable crops, China, Intercropping, Sustainable agriculture

4. Chapter I

INTRODUCTION

It may be referred to as an ancient and traditional cropping system, but has a serious potential to contribute to a modern and sustainable agriculture in China. Intercropping, defined as a kind of multiple cropping system with two or more crops grown simultaneously in alternate rows in the same area (Federer, 1993) while minimizing competition, is practiced in China for thousands of years. In general, intercropping can be done with field, vegetables and even tree crops. Available growth resources, such as light, water and nutrients are more completely absorbed and converted to crop biomass by the intercrop as a result of differences in competitive ability for growth factors between intercrop components. The more efficient utilization of growth resources may lead to yield advantages and increased stability compared to sole cropping and, hence, offers an option for a sustainable low-input cropping system and economic benefits. Though this practice may have some drawbacks, some may be overcome by proper intercrop selection and management. Table 1 lists possible benefits and uncertainties of intercropping compared to classical monocropping systems.

In China, estimations between more than 28 million ha (Li et al., 2007) and 3.4×10^7 ha (Li, 2001) of annually sown area are under intercropping, with a big share in agroforestry (Table 2).

Not only tradition and extent of intercropping practice in China are liable for the appearance of a great number of studies dealing with intercropping in China, but the alternatives and options of this cropping system for ecological and sustainable agriculture, especially over the last decade. Foremost in the last decade, Chinese researchers showed an increasing interest in this cropping system. However, research on intercropping has mostly been carried out in Africa, India and Australia, leading to a better understanding of these systems in these countries.

Table 1: Benefits and uncertainties of intercropping systems

Benefits	Uncertainty
Maximized land use	Limited possibilities for production mechanization
Implement more than one harvest per year (e.g. with relay intercropping)	Harvesting produce more difficult
Diversification of crops for market supply	Higher management demand
Risks of crop failure may be reduced	No extensive production of stable or cash crops
Farmers may be better able to cope with price variability	A poorly chosen intercrop competes with main crop
Higher yield and improved resource efficiency	Intercropping may not significantly improve the soil nitrogen levels
Boost the soil nitrogen content in the medium to long term especially when legumes are involved	Herbicide use may be constrained
Soil structure may improve if plants with various root structures are grown	
Rotation effect and improving soil erosion control	
Reducing pests and weeds	
Reducing reliance on energy-intensive farming inputs and therefore less eutrophication and emission	

Internationally published studies dealing with intercropping convey the impression that the main interest of these systems lies upon plant nutrition as most of them consider the increased nutrient availability and uptake, the soil nitrate content and N leaching under intercropping in comparison with monocropping (Table 3). However, there is still a research gap considering crop production and cropping designs, especially as the studies about nutrient supply and availability in intercropping systems are mostly conducted under controlled conditions. The performance and behavior of intercropping systems in comparison to monocropping systems under field conditions are still fairly unknown. In addition, other aspects like N-efficiency, water-use efficiency, influence of tillage, pests and diseases or even the calculation of land equivalent ratios (LERs) have not been considered in the international literature so far. Further, in contrast to other countries like India or Africa, in China there is no special breeding of varieties suitable for intercropping comparable to a few approaches in some African regions. In rural areas, intercropping is practiced as a so-called unconscious cropping system, which forms a big part of the whole

Table 2: Average farm size, intercropping practice and arable land in Africa, China and India*

Country	Farm size (ha)	Intercropping area	Intercropped species	Arable land (million ha) (of total land)	Population (million)
Africa	2	83 % of all cropped land in Northern Nigeria; 94 % in Malawi	Cowpea, cassava, plantain, yam, rice, sorghum, millet, maize, sweet potato, okra, cocoa, soybean, chickpea, pigeon pea, peanut, beans	182.3 (6 %)	812.6
China	0.1	20 – 25 % of arable land	maize, soybean, peanut, potato, wheat, millet, faba bean tobacco, cotton, sorghum, sesame, garlic, vegetables, cassava	137.1 (16 %)	1 320.9
India	1.2 - 2.7	17 % of arable land	peanut, pigeon pea, maize, soybean, sugarcane, jatropha, rubber, cabbage, coconut, banana, cassava, sorghum, rice, mustard, amaranth, potato, wheat	160.6 (57 %)	1 081.3

* Source: American Society of Agronomy, 1976; Beets, 1982; Cohen, 1988; FAO, 2004; Li, 2001; Li et al., 2007, Vandermeer, 1989, Wubs et al., 2005.

intercropping area in China: monocropped fields that are enclosed with one row of a different crop to separate them from neighbouring fields, limited field size turning the borderlines of one field to another to an intercropping pattern (Fig. 5). Of course, intercropping is common and widespread, but the main reason for a farmer to carry out intercropping is to use all available land for production, as arable land is scarce. Due to restricted land-use rights Chinese farmers are not able to increase the size of their farms and expand cropping areas. Hence, maximizing yields is only possible by optimizing crop management strategies leading to a better utilization of natural resources over space and time. Intercropping may be a suitable strategy to do so as multiple crops can be grown simultaneously over space and time offering the chance to better utilize solar radiation, nutrients and water over the growing period. Intercropping bears more advantages and is more than maximized field exploitation.

Table 3: Overview over experiments and main researches dealing with intercropping cereals in China

System	Main research	Reference
wheat/maize	Spatial compatibility and temporal differentiation of root distribution	Li and Zhang, 2006; Li et al., 2006.
	Competition-recovery production principle	Li et al., 2001b; Zhang and Li, 2003.
	NO_3^- content in the soil profile or NO_3^- concentration in the rhizosphere or N uptake	Li et al., 2005; Song et al., 2007; Ye et al., 2005 ; Zhang and Li, 2003.
	Inner- and border-row effects	Li et al., 2001a.
wheat/soybean	N_2 fixation rate and/or N uptake	Li and Zhang, 2006; Li et al., 2001b; Zhang and Li, 2003.
	Competition-recovery production principle	Li et al., 2001b.
	Inner- and border-row effects	Li et al., 2001a.
wheat/faba bean or maize/faba bean or wheat/chickpea or maize/chickpea	Utilization, availability and uptake of P	Li and Zhang, 2006; Li et al., 2003; Li et al., 2004; Li et al., 2004; Li et al., 2007; Zhang and Li, 2003; Zhang et al., 2004.
	Utilization, availability and uptake of N	Li et al., 2003; Song et al., 2007; Xiao et al., 2004; Yu and Li, 2007; Zhang et al., 2004.
	NO_3^- content in the soil profile	Li et al., 2005.
	Spatial compatibility and temporal differentiation of root distribution	Li et al., 2006.
maize/peanut	Reduction of Fe chlorosis	Inal et al., 2007; Zhang and Li, 2003; Zhang et al., 2004; Zheng et al., 2003; Zuo et al., 2004.
	N_2 fixation rate and/or N uptake	Zuo et al., 2004.
	Nutrient (Fe, P, N, K, Ca, Zn, Mn) supply	Inal et al., 2007.

In China, two main systems of intercropping are common: strip intercropping and relay intercropping (Fig. 1, 3). Strip intercropping is defined as cropping two or more crops simultaneously in different strips, wide enough to permit independent cultivation but narrow enough for the crops to interact agronomically (Vandermeer, 1989). Relay intercropping means the maturing annual crop is interplanted with seedlings or seeds of the following crop (Federer, 1993). Intercropping can be practiced further as an additive or a replacement system (Fig. 2). Additive means that one crop is planted in a similar arrangement to its sole-crop equivalent, and a second crop is added, so that the total plant

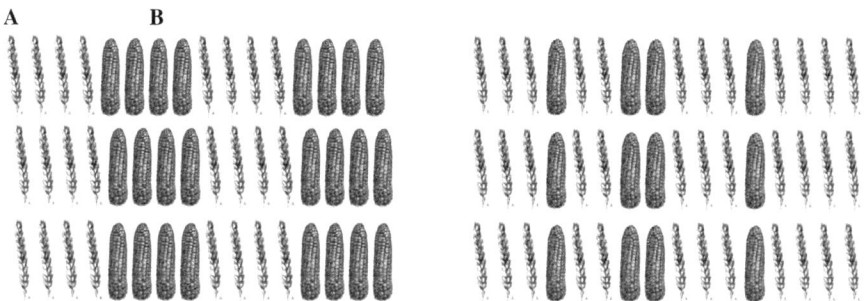

Figure 1: Scheme of the two main intercropping systems in China: Strip intercropping (A), where two or more crops are grown simultaneously on the same field in different strips, e.g. four rows of crop A and four crops of crop B. Relay intercropping (B), where the maturing annual plant is interplanted with seeds of the following crop, e.g. 75% of wheat is sown in autumn and a few days or weeks before wheat harvest, maize is interplanted

density increases (Keating and Carberry, 1993). In contrast, a number of a few rows of one crop can be replaced by a second crop with total plant density not necessarily changing.

Two practical examples, first for strip or row intercropping, second for relay intercropping, should be mentioned which show the positive implications of intercropping. First, in China's northeast 0.3 × 106 ha of maize fields have been converted to intercropping with sweetclover. Even by planting one-third of the field with sweetclover, the results in maize yields are about the same as those from monoculture. But there are additional 15 t ha^{-1} of sweetclover to feed three cows a year (Wen et al., 1992). Thus, without maize yield reduction, the amount and quality of pasture can be improved leading to an extension or an increase in livestock, as livestock farming is often based upon substrate fodder. Especially

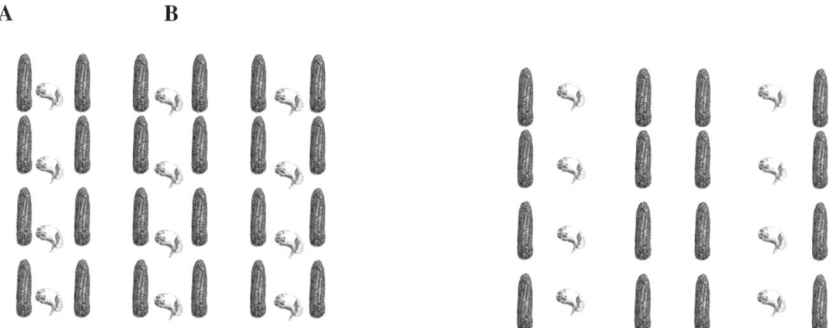

Figure 2: Considering plant density, two main systems of intercropping exist. (A) In an additive intercropping system, crop A is planted in a similar arrangement and amount to its sole-cropped equivalent, and a crop B is added. Total plant density increases. (B) In a replacement system, a few rows of crop A are replaced by crop B. Total plant density does not necessarily change

around the big cities Beijing, Tianjin and Shanghai and in the provinces Jiangsu, Zhejing and Fujian, cattle are mostly used for unimproved production and draught (Guohua and Peel, 1991). Improved production for milking is necessary to increase in nearly all provinces, and therefore, good and enough pasture is a big point. Second, relay intercropping showed that it is possible to increase grain yields of summer maize without a decrease in winter wheat productivity. This is remarkable, since only 75% of winter wheat sowing acreage is cultivated (Böning-Zilkens, 2004). But there is additional summer maize yield because of an elongated growing season.

A   B

Figure 3: Two examples of typical intercropping systems in the North China Plain: (A) Strip intercropping of maize and soybean (l.), where two or more crops are grown simultaneously on the same field in different strips. Three rows of soybean alternate with maize. These small strips often mark the borderline between two fields owned by different farmers and are the so-called unconscious intercropping. (B) Relay intercropping of wheat and maize, where the maturing annual plant is interplanted with seeds of the following crop (r.). Wheat and maize have a few days or weeks of overlapping growing season. When the wheat is harvested, the maize still grows for more than 3 months. In such a system plant density of maize is less than in monocropped systems (pictures: Zhang, F. and Feike, T.)

GENERAL QUESTIONS ABOUT COMPETITION AND FACILITATION

A brief basic introduction on intercropping, the influencing factors like competition for radiation, CO_2, water and nutrient availability, the ratio of competition and facilitation, resource capture and conversion efficiency is given by Vandermeer (1989) and Fukai and Trenbath (1993). Previous studies mainly dealt with competition (Keating and Carberry, 1993; Tsubo et al., 2001) and less with facilitation. However, interspecific competition and facilitation are two aspects of the same interaction, turning the system of intercropping to a successful one under some circumstances. Jolliffe (1997) pointed out that mixtures are significantly more productive than pure stands on an average of 12%. Current studies pronounce the interspecific facilitation (Li et al., 2001a,b, 2007; Zhang and Li, 2003) and hence, cultures or cereals suiting to each other for a better cropping management. Factors like root distribution (Li et al., 2006), root interactions (Inal et al., 2007; Li et al., 2003b; Zhang et al., 2004; Zheng et al., 2003), above- and below-ground interactions meaning row effects (Li et al., 2001a; Zhang and Li, 2003), N, P, K (Li and Zhang, 2006; Li et al., 2003b, 2004, 2007) and Fe (Zheng et al., 2003; Zuo et al., 2004) supply or modeling competition (Bauer, 2002; Kiniry et al., 1992; Piepho, 1995; Rossiter and Riha, 1999; Yokozawa and Hara, 1992) have been studied recently.

Competition or facilitation extents are difficult to class with their overall extent because of their intermingling. The success of intercropping is attempted to be measured by calculating the relative performance of the species. The most common parameter for judging the effect of intercropping is the land equivalent ratio (LER). A long, detailed and comprehensive list of international studies using the LER as an indicator of success is shown by Innis (1997). There, in almost all studies, LER was greater than 1 indicating that the intercropped species overyielded its monocropped counterparts. The LER is defined as

$$LER = \sum_{j=1}^{n} Y_{ji} / Y_{js}$$ (Wubs et al., 2005).

Y_{ji} = yield of component crop j in intercropping
Y_{js} = yield of component crop j in sole cropping

Jolliffe (1997) promoted the Relative Yield Ratio (RYT) instead of the LER, but the RYT may be equivalent to LER as the same formula is used:

$$RYT = [(Y_i)_m / (Y_i)_p] + [(Y_j)_m / (Y_j)_p]$$ (Jolliffe, 1997).

Y = yield
i, j = species 1 and 2
m = species mixture
p = pure stand

RYT and LER do not express the simple ratio of mixture to pure stands, nor do they involve equal populations and areas allocated to mixture and pure stands (Jolliffe, 1997). Instead, Jolliffe calculated the Relative Land Output (RLO):

$$RLO = (Y_i + Y_j + \ldots)_m / (Y_i + Y_j + \ldots)_p$$ (Jolliffe, 1997).

In order to get some information about the competitive ability of one species to another, the aggressivity (A) can be calculated:

$$A_{ab} = [Y_{ia} / (Y_{sa} * F_a)] - [Y_{ib} / (Y_{sb} * F_b)]$$ (Li et al., 2001a).

Y = yield
s = sole cropping
i = intercropping
F = proportion of the area occupied by the crops in intercropping
a, b = crop 1 and 2

When $A_{ab} > 0$, the competitive ability of crop A exceeds that of crop B. In addition, the Nutrient competitive ratio (CR) is given in the following equation:

$CR_{ab} = [NU_{ia} / (NU_{sa} * F_a)] / [NU_{ib} (NU_{sa} * F_b)]$ (Li et al., 2001a).

NU = nutrient uptakes by species

When $CR_{ab} > 1$, the competitive ability in taking up nutrients of crop A is more efficient than that of crop B. At least, the cumulative Relative Efficiency Index (REI_c) is a measure that compares the proportional change in total dry matter within a given time interval of one species relative to another:

$REI_c = K_{crop\,a} / K_{crop\,b}$ (Hauggaard-Nielsen et al., 2006).

$K_{crop\,ab}$ = dry matter$_{ab}$ at time t_2 / dry matter$_{ab}$ at time t_1

A TRADITIONAL CROPPING SYSTEM AS A CONTRIBUTION TO SUSTAINABLE AGRICULTURE IN CHINA

Historically, intercropping has already been proven for the Dong Zhou and Qin dynasties (770–206 BC) as a special form of crop rotation. Initiated from cropping of forests together with grains or cereals, the intercropping practice went further on with hemp, soybean, mung bean, rice and cotton as well as a system of intercropping of grains with green manure plants.

One of the early and important written documents about crop rotation and intercropping as a sub-item was the "Important Means of Subsistence for Common People", dated back to the period of Wei and Jin dynasties (200–580 AD). It pointed out the possibility to improve soils by multiple cropping of red bean, mung bean and flax and described the theoretical and technical basis for proper rotations of leguminous plants and cereals. During the Ming Dynasty (1368–1644), the "Complete Works of Agronomy" was written by the agronomist Xu Guangqi, who summarized his experience on intercropping of wheat and broad bean (Gong et al., 2000). Besides, multi-component systems with some pertaining to intercropping are often reported in ancient literature, e.g. "Essential Farming Skills of the People of Qi" (~600 AD), "Agricultural Treatise of Chen Fu" (1149) and "Complete Treatise on Agriculture" (~1600) (Li, 2001).

In ancient Chinese agronomy literature it is postulated that farming activities should be in accordance with seasons, climate, soil conditions and nutrient input. Ellis and Wang (1997) showed in their regional field study on traditionally cultivated and used agro-ecosystems in the Tai Lake Region that these systems were capable of sustaining high productivity for more than nine centuries. Further, Chinese philosophers always pronounced the harmonious relationship between humans, nature and environment, e.g. Zhou Yi's famous treatise "The Book of Change" or the Yin and Yang theory (Li, 1990), which had profound influence on the practice and formulation of integrated farming systems in modern Chinese agriculture policies, e.g. China's Agenda 21. This Agenda, approved on 23 March 1994 in the form of the White Paper on China's Population,

Environment and Development, lays down some basic principles for the comprehensive management of a sustainable agricultural development. As a consequence, Chinese Ecological Agriculture, not to confuse with European Organic Farming, puts a great deal of emphasis not only on the protection of the environment and the saving of resources but on the all-around development of the rural economy and specifically on rural income-generating activities (Sanders, 2000). Intercropping might be one suitable strategy.

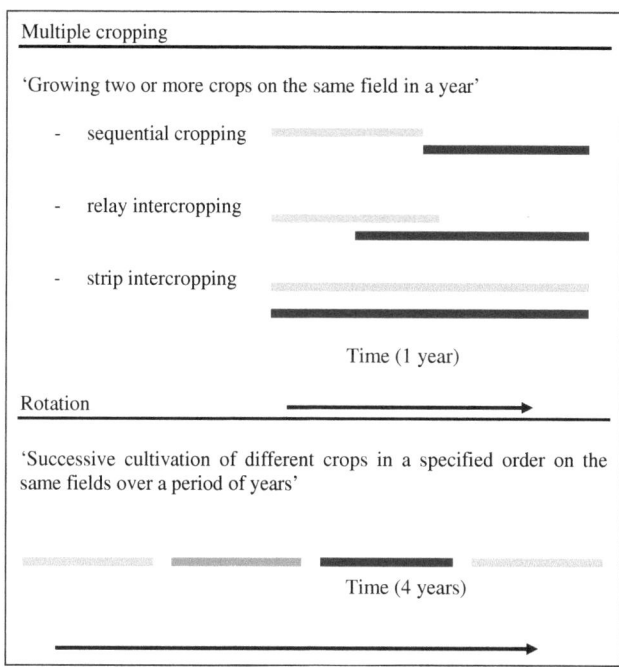

Figure 4: Differences between and definitions of rotation, multiple cropping and intercropping (Graphic taken from Wageningen University, presentation 2002, definition of rotation added by the author)

In a public western view, modern agricultural development, except some European organic farming labels, has less to do with ancient philosophy, but more with intensive monocultures (cash crop), extensive use of energy, fertilizer and pesticides and machinery replacing manpower and human labour force. Considering the actual agricultural situation in China (Lin, 1998; Lüth and Preusse, 2007) – production as well as markets – intercropping as a well-adapted cropping system in this country is an option to solve the massive environmental problems caused by high use of fossil energy-based inputs and a non-resource-preserving agriculture. Hence, a traditional cropping system could turn out to be a modern one (Lu et al., 2003; Zhen et al., 2005).

Intercropping is known as a system being more efficient in poorer soil and environmental conditions – because of a higher uptake and utilizing efficiency of resources like nutrients – and low-input cultivation, but losing this advantage in high-input cultivation. A theory about intercropping and input level says that the productivity of intercropping systems is higher than for pure crop situations when the input is low, but that this advantage decreases as the inputs increase (Wubs et al., 2005). Besides, seriously managed intercropping grants an option for a sustainable low-input cropping system with some kind of rotation effect, reducing pests and weeds, reducing reliance on energy-intensive farming inputs and therefore less eutrophication and emission, improving soil erosion control and, after all, giving economic benefits.

The enormous increase in grain yield and production per capita in the last 50 years in China appeared mostly due to the increased industrial energy inputs, especially in the form of chemical fertilizer (Tong et al., 2003) and irrigation water (Binder et al., 2007). It is not only to increase yield, but to compensate for the loss and degradation of the best lands through industrialization, erosion and soil misuse because of excessive irrigation, unadjusted cropping systems and chemical fertilizer.

For example, in the middle reaches of Heihe River in the Hexi Corridor region, the change of cultivation modes – crop–grass intercropping instead of monocropping – intimated a reduction in soil wind erosion and a halt in sand entrainment (Su et al., 2004). In this region it occurs that dust transport from farmlands is about 4.8–6.0 million tons per year and consequently higher than that of sandy desert dust transport in the same region.

In entire China, the use of mineral fertilizers grew more than 50 times from 1962 (0.63 million tons) to 1994 (33.18 million tons) with 80% being N fertilizer (Inal et al., 2007). The average fertilizer consumption in 2002 was 277.7 kg ha^{-1} arable land (FAO, 2006), rising in the irrigated areas to 450 kg ha^{-1}. Considering N-fertilization amounts, China ranks first in the world. Simultaneously, the yield per unit chemical fertilizer use decreased from 164 (rice), 44 (wheat) and 93 (maize) kg kg^{-1} in 1961 to 10 (rice), 6 (wheat) and 9 (maize) kg kg^{-1} in 1998 (Tong et al., 2003). This decline appears as a result of fertilizer saturation, soil degradation, strong soil, atmosphere and water pollution and poor and low-quality land use. Further, more than 100,000 people were poisoned by pesticides and fertilizers during 1992–1993, and more than 14,000 of them died (Sanders, 2000). Thus, low input cropping systems sometimes yielding (economically) as high as high-input cropping systems are to be favored, further offering the change to save natural and ecological resources.

China is such a large country that any land use (Xiaofang, 1990) change would contribute greatly to changes in the global world (Tong et al., 2003). The country is known as an important source of methane from rice paddies as well as atmospheric nitrous oxide generated by the increasing use of large amounts of low grade and highly volatile ammonium bicarbonate fertilizers. In contrast, China has an urgent demand for food for a rapidly growing population (Gale, 2002; Lu and Kersten, 2005; Ministry of Agriculture of the People's Republic of China 2004). Only 25.7% of Chinese agricultural area is suitable for arable land usage, with 2.2% being permanent crops (FAO, 2006). With 1.32 million people, China has 20% of the world's population but only 9% of the world's arable cropland. There is only 0.1 ha per capita, which is one-third of the world's average (Wen et al., 1992). Following Tong et al. (2003), more than 60% of all cultivated land is poor in nutrients, and only 5–10% is free from drought, water logging or salinity. Two percent of China's total land area can be considered desertified by human-induced resource degradation (Sanders, 2000). This shows that China has a great demand for resource and environment-saving agriculture systems being able to feed the country. Intercropping as a widely practiced and accepted cropping system could contribute to a more sustainable land use in this context.

In respect of the enormous environmental problems, like water shortage and pollution, over-fertilization, high nitrous emissions, leaching, soil degradation, erosion, etc., caused by agriculture in China, sustainability is an increasing factor to consider in future preventing natural catastrophes and preserving production levels (Lei, 2005). Climate change and resulting implications for sustainable development was a topic in the Regional Implementation Meeting of the United Nations Economic and Social Commission for Asia and the Pacific, held in November 2007. Around 15.5% of the GDP in 2006/2007 in China came from agriculture, animal husbandry and forestry, though more than 50% of the population depends on agriculture for their livelihoods and hence, a large proportion of the population depends on the climate and the climate change (Prabhakar, 2007). For 2050, the UN figures on scarcity of freshwater availability affecting more than a billion people in Central, South, East and South-East Asia. Additionally, crop yields are predicted to decline in parts of Asia between 2020 and 2050 about 2.5 to 30%. Freshwater availability, droughts and floods due to greenhouse gas emissions, water pollution and soil degradation will be the main problems. Therefore, natural resource and integrated ecosystem management are identified to be major priority actions. The Commission accentuated clearly (Prabhakar, 2007): "Revisiting the existing cropping patterns and systems is needed". As monocropping means higher risk, in terms of income security, nutritional diversity in rural areas and the possibility of severe impacts to large areas due to pest and disease outbreak in a changing climate, mixed and intercropping practices are the only alternative that may have multiple benefits.

THE NATURE AND EXTENT OF CHINESE INTERCROPPING

One-third of China's cultivated land area is used for multiple cropping (Fig. 4), and a half of the total grain yield is produced with multiple cropping (Zhang and Li, 2003). At present, about more than 70% of farm products are attributed to the improved multiple cropping systems like rotations or intercropping (Zhang et al., 2004). Between 1949 and 1995, the multiple cropping index, meaning the sown area:arable area ratio, increased from 128% to 158%, according to an increase of 2.7×10^7 ha of farmland (Li, 2001). Although, intercropping is only one example of the various aspects of multiple cropping, it is substantial as China's intercropping area is the largest in the world. As an example, in 1995 the area under wheat intercropped with maize was about 75,100 ha in Ningxia, producing 43% of total grain yield for the area (Li et al., 2001a). While the most common agricultural land use in the Heihe River Basin before 1980 was to crop wheat, the intercropping of wheat together with maize increased after 1980. Today, 20% of the agricultural land in this region is sown with wheat, 40% with maize and 40% with wheat intercropped with maize (Yamazaki et al., 2005). Furthermore, intercropping has become the most common cropping system for peanut production in northern China (Zuo et al., 2004). Of 31 provinces 27 have land under peanut production. In 1991, the UNEP granted the Zhang Zhuang region the Global 500 award, given for sustainable farming systems in the sense of Chinese Ecological Agriculture (Sanders, 2000). In this region there were altogether ten different modes of intercropping, including cereals and oil crops and cereals and vegetables.

The most common intercropping types considering cereals or grains are those of wheat and maize, wheat and cotton, wheat and faba bean, wheat and soybean, maize and soybean, maize and faba bean and maize and peanut. Sorghum and maize are often used to enclose fields. In addition, there is a vast range of possible combinations of grains together with vegetables. Rice is mostly cultivated in the south where most of the high-quality land is found. In contrast, wheat is cultivated in central China and in the north, and maize is cultivated in central, the northeast and the north of China. The center of cereal production moved slightly towards northern China (Tong et al., 2003), with the North China Plain being China's granary. Various inter- or relay-cropping patterns are practiced mainly in the north, the northeast and northwest and the southwest, especially in Xinjiang, the corridor in the Gansu, Yinchuan plain, Hetao in Inner Mongolia, the Northeast Plain, the North China Plain, along the Yangtze and Yellow rivers, and also in lower dry lands and in hilly areas of South China (Ren, 2005). Wheat intercropped with maize has become increasingly popular in the irrigated area of the Hexi Corridor in the Gansu province, along the Huanghe River in Ningxia and in Inner Mongolia regions (Zhang and Li, 2003). In China's northwest, wheat/soybean and maize/faba bean intercropping systems are well established, and peanut/maize intercropping is

Figure 5: Average field size in China is very small, so the collectivity of field borders can be considered as intercropping in a larger scale (pictures: Feike, T.). Typical sequential borderlines in the North China Plain are between soybean, cotton, maize or sorghum, various vegetables, especially cabbage, and interjacent poplar trees

widespread in the northern parts of the country (Zhang and Li, 2003). In the southwest, wheat–maize intercropping predominates within irrigated spring maize cropping systems. Also within rainfed spring maize cropping systems, intercropping is common. Especially in Sichuan province, wheat–maize intercropping is the most common agroecosystem model (Meng et al., 2006) (Fig. 6).

INTERCROPPING TYPES AND REGIONS

There are different approaches to divide China's agricultural land into specific agricultural regions. The most common is to partition the country into nine, respectively ten, agricultural regions (Guohua and Peel, 1991) situated in middle and eastern parts of China, depending on percentage of cultivation, climatic features and production systems. The agricultural region boundaries are not necessarily authoritative. The agriculture production system zone code of the PR China actually splits the country into 12 zones (FAO, 2007). As the title already tells, the production system – whether it is single or multiple cropping, cropping for uplands or paddy field – is decisive. Meng et al. (2006) analyzed the various cropping systems and potential production with regard to maize production, and divided China's cultivated land into six agricultural regions depending on significant differences in maize cropping patterns and practices.

Concerning intercropping, China's cultivated land can be classified into four main types and regions – illustrated in Fig. 6 – which is just a first and general approach to classify intercropping regions at all because there has been no detailed statistic or documentation dealing with intercropping in particular so far. This basic classification takes the underlying potential for intercropping and the intercropping practice into account (Atlas of the PR of China, 1989; Meng et al., 2006). In general, the agricultural regions in the north, northeast and northwest have more potential for intercropping than the regions in the south, even if the southeast has the highest average precipitation per year, a subtropical, humid and monsoonal climate and with around 135–242%, the highest multiple cropping index in China. The southwest is a very important agricultural region for China with a great diversity of crops, fruits and vegetables, but the climate conditions allow more flexible rotations, thus replacing intercropping systems. Going from north to south the cropping systems change from one crop a year with a great potential for intercropping to relay intercropping of especially maize and wheat and double cropping systems and at least three cropping seasons per year with different kinds of rotations and increasing number of paddy fields (Table 4).

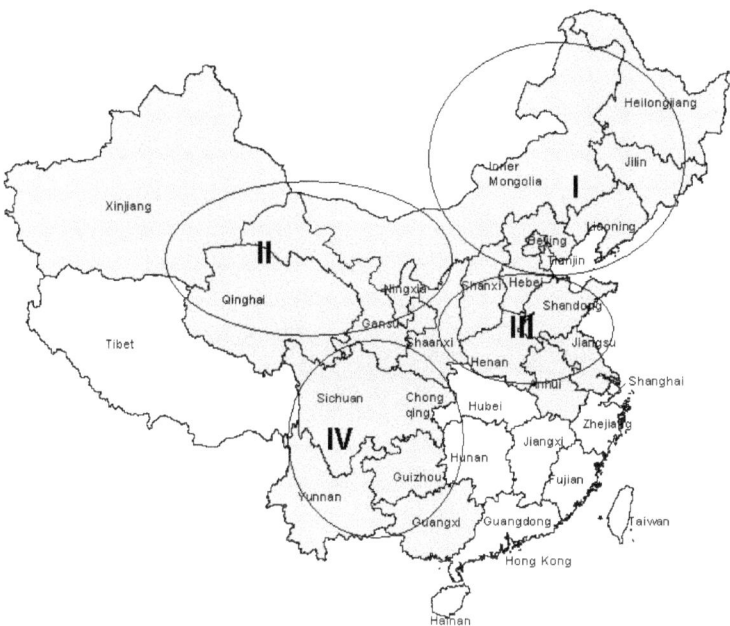

Figure 6: Provinces where intercropping with cereals is popular (*grey*); regions where intercropping with cereals is less common (*white*) (Graphic taken from USDA, 2007, marking of intercropping regions added by the author). In a rough and simplified visualization, China can be classified into four intercropping regions I to IV: Going from Northeast and North (I) to Northwest (II) and Yellow-Huai River Valley (III) and finally to Southwest (IV) the cropping systems change from one crop a year (I + II) with a great potential for intercropping to relay intercropping (III) of especially maize and wheat and double cropping systems and at least three cropping seasons per year (IV) with different kinds of rotations, and rotations replacing intercropping

Type I: Single Cropping with Great Intercropping Potential

The Northeast is characterized by a cold-temperate/semi-humid and temperate/humid climate zone in the north of the agricultural zone and temperate continental monsoonal climate zone. But there are also temperate, humid, monsoonal to subtropical, semi-humid, monsoonal climate conditions to be found.

In the North, there is a warm-temperate, semi-humid, monsoonal and also temperate continental climate. Dark brown soils, phaeozems, chernozem-castanozem-dark loessial soils, brown soils and cinnamon soils are predominant with high accumulation of organic matter. Especially in the Northeast, the soils tend to be slightly acidic or calcareous, with the latter being a problem for, e.g., peanut production. In calcareous soils, the Fe availability is weak because of immobilization of Fe in alkaline soils. Peanut is an important crop in this region, and the disadvantage of Fe availability

Table 4: Intercropping types and regions and their characteristics and potentials* in China

Agricultural region	Cropping conditions and characteristics	Intercropped species	Type of production
Northeast (NE) and North (N): • Liaoning • Jinlin • Heilongjiang • Parts of Inner Mongolia • Beijing • Tianjin • Parts of Hebei • Parts of Shanxi	*Climatic features*: *middle to warm temperate zone* > *10°C accumulated temperature (°C)*: NE: 1,300–3,700 N: 200–3,600 *Average temperature (°C)*: −12 to −14 *Sunshine (hours)*: 2,300–3,200 *Frost-free period (days)*: 100–200 *Rainfall (mm/year)*: NE: 500–800 N: 200–600 *Altitude (m)*: 50–100 *Soils*: *siallitic and calcareous soils predominant* Dark brown soil and phaeozem zone; chernozem-castanozem-dark loessial soil zone; brown soil and cinnamon soil zone *Multiple cropping index (%)*: 0–135	Maize/** soybean Maize/peanut Maize/potato Wheat/broomcorn millet	*TYPE I: single cropping with great intercropping potential* • crops: maize, soybean, spring wheat, rice, sorghum, millet, sesame, potato, sugar beet, flax, peanut, ambary hemp, cotton • yield and production level of cereals: 5,155 kg ha^{-1} • irrigated area (1,000 ha): 1765.2 • consumption of chemical fertilizer (10,000 tons): 118.2

Table 4 (continued)

Agricultural region	Cropping conditions and characteristics	Intercropped species	Type of production
Northwest: • Gansu • Qinghai • Nigxia • Xinjiang • Parts of Inner Mongolia • Parts of Shaanxi	*Climatic features: cold temperate to subtropical zone* *> 10°C accumulated temperature (°C):* 2,000–4,500 *Average temperature (°C):* 0–12 *Sunshine (hours):* 2,600–3,400 *Frost-free period (days):* 140–170 *Rainfall (mm/year):* 10–250 *Altitude (m):* 300–3,000 *Soils: gypsum-bearing and calcareous soils predominant* subalpine meadow soil zone; brown desert soil zone; grey dessert soil zone; sierozem-brown calcic soil zone *Multiple cropping index:* 0–135	Maize/potato Maize/bean Wheat/maize Wheat/buckwheat Wheat/millet Wheat/tobacco Wheat/soybean	<u>TYPE II</u>: *single cropping for cold climate and semi-arid crops to double cropping for irrigation farming* • crops: maize, wheat, millet, broomcorn millet, oats, buckwheat, potato, highland barley, sorghum, rice, rape, soybeans, sugar beet, cotton, flax, hemp, peanut, pea, broad bean • yield and production level of cereals: 4,228 kg ha^{-1} • irrigated area (1,000 ha): 1438.2 • consumption of chemical fertilizer (10 000 tons): 75.6

Table 4 (continued)

Agricultural region	Cropping conditions and characteristics	Intercropped species	Type of production
Yellow-Huai River Valley: • Parts of Hebei • Parts of Shanxi • Shandong • Henan • Parts of Shaanxi • Parts of Anhui • Parts of Jiangsu	*Climatic features: warm temperate to subtropical zone* > *10°C accumulated temperature (°C):* 3,400–4,700 *Average temperature (°C):* 10–14 *Sunshine (hours):* 2,200–2,800 *Frost-free period (days):* 170–220 *Rainfall (mm/year):* 500–1,100 *Altitude (m):* 50–100 *Soils: calcareous soils predominant Brown soil and cinnamon soil zone; yellow brown soil and yellow cinnamon soil zone* *Multiple cropping index:* 1–190	Wheat/maize Wheat/cotton Maize/soybean in rotation with wheat Wheat/garlic in rotation with maize	*TYPE III: double cropping with potential for relay intercropping* • crops: maize, wheat, soybean, peanut, cotton, vegetable, millet, potato, sorghum, sweet potato, rice, sesame, ambary hemp, tobacco, pea, sugarcane • yield and production level of cereals: 4,876 kg ha^{-1} • irrigated area (1,000 ha): 3369.2 • consumption of chemical fertilizer (10,000 tons): 297.9

Table 4 (continued)

Agricultural region	Cropping conditions and characteristics	Intercropped species	Type of production
Southwest: • Parts of Guangxi • Sichuan • Chongqing • Guizhou • Yunnan • Parts of Shaanxi	*Climatic features*: subtropical to tropical zone *> 10°C accumulated temperature (°C)*: 3,500–6,500 *Average temperature (°C)*: 15–18 *Sunshine (hours)*: 1,200–2,600 *Frost-free period (days)*: 240–360 *Rainfall (mm/year)*: 800–1,600 *Altitude (m)*: 200–3,000 *Soils*: *ferrallitic and ferro-siallitic soils predominant* Alpine meadow soil zone; subalpine meadow soil zone; read earth and yellow soil zone; lateritic red soil zone *Multiple cropping index*: 135–242	Maize/beans (sorghum) in rotation with wheat Maize/potato/wheat Wheat in rotation with maize/soybean (green bean) Maize/beans in rotation with wheat Maize/wheat Wheat in rotation with maize/sesame Wheat/vegetable in rotation with sweet potato/maize/soybean Maize/sorghum in rotation with wheat Maize in rotation with maize/sweet potato Maize/potato in rotation with wheat Rice/wheat Rice/rape Potato/maize Maize/cassava in rotation with soybean Maize/soybean in rotation with sunflower Wheat in rotation with maize/sesame-maize/potato Vegetable in rotation with maize/sweet potato Maize/sweet potato in rotation with wheat/vegetable Rape/maize Wheat/vegetable in rotation with maize/vegetable	<u>*TYPE IV*: *three cropping seasons per year with rotations replacing intercropping*</u> • crops: rice, maize, wheat, sweet potato, sorghum, rapeseed, sugarcane, peanut, tea, cotton, ambary hemp, tobacco, millet, cassava, soybean, pea • yield and production level of cereals: 4,605 kg ha^{-1} • irrigated area (1,000 ha): 1349.2 • consumption of chemical fertilizer (10,000 tons): 140.3

* Source: Atlas of the PR of China (1989); Meng et al. (2006); National Bureau of Statistics of the Peoples Republic of China, 2006; The National Physical Atlas of China = Chinese Academy of Sciences (1999).
** meaning "intercropping".

can be remedied through intercropping as shown later.

In both, Northeast and North of China, with the provinces and municipalities Liaoning, Jinlin, Heilongjiang, parts of Inner Mongolia, Beijing, Tianjin, parts of Hebei and Shanxi, intercropping is widespread. The climatic features allow only one crop per year. The average precipitation is between 200 and 800 mm year^{-1} with rain falling mainly in summer, whereas spring droughts are frequent. The winters are long and cold, and there is a large daytime–nighttime temperature gap during the whole growing season. In addition, the average temperature is very low. The varieties grown in this region have to be fast-maturing varieties. Only crops which prefer semi-moist and warm conditions can be grown, thus reducing the cultivation range to especially maize, soybean, peanut, potato, spring wheat and millet. Generally, the production system is based upon rainfed conditions. In Hebei, Shanxi and around Beijing, irrigation is also practiced. In the northeast as well as in the north, intercropping can provide optimal site utilization, a higher yield compared to monocropping and an improved diet diversification.

Intercropping Maize with Peanut

Peanut is the major oilseed crop in China constituting 30% of the land's total oilseed production and 30% of the cropped area (Zhang and Li, 2003). But especially in north China with its calcareous soils, iron deficiency chlorosis is often observed and Fe deficiency is one of the most common yield-limiting nutrients that causes serious economic problems in peanut monocropping systems (Zuo et al., 2004). Maize has a great potential to improve the Fe nutrition of peanut within an intercropping system by rhizosphere interactions (Fig. 7). In an experiment of Zhang and Li (2003), young leaves of peanut plants in rows 1–3 from the maize grew without visible symptoms of Fe deficiency, while those in rows 5–10 showed a variable degree of chlorosis. They were all chlorotic when roots were separated. Peanut is a strategy I, and maize a strategy II species. Strategy I plants respond to Fe deficiency with increased ferric reductase activity of roots and acidification of the rhizosphere by releasing protons from the roots. Strategy II plants excrete phytosiderophores into the rhizosphere thus being more efficient in Fe deficiency surroundings. They mobilize Fe (III) and benefit the iron nutrition of maize as well as peanut (Marschner, 1986).

The results indicate the importance of intercropping systems as a promising management practice to alleviate Fe deficiency stress (Inal et al., 2007), because soil amendments and foliar application of Fe fertilizers are usually ineffective or uneconomical for correcting Fe chlorosis.

Figure 7: Intercropping of maize and peanut reduces iron chlorosis in peanuts on calcareous soils (pictures: Zhang, F.). (A/B): Differences between (strip) intercropped (l.) and monocropped (r.) peanut in the field. (C/D): Differences between intercropped (l.) and monocropped peanut (r.) in a pot experiment. In the pot experiment as well as in the fields, peanut shows less Fe chlorosis when intercropped with maize

But in addition and especially in calcareous soils, the effects of soil moisture on soil iron availability under intercropping could be more complicated compared to monocropping. Zheng et al. (2003) showed that the Fe nutrition of peanut intercropped with maize could be affected by soil moisture condition. Root growth of peanut was significantly inhibited at 25% soil water content compared to those at 15% soil water content. Also, chlorophyll content in the new leaves of intercropped peanut decreased and leaves became chlorotic at 25% soil water content.

The improved Fe availability is the underlying reason for an increased N uptake. Competitive interactions between maize and peanut for N and improvement of Fe uptake by peanut were likely to be important factors affecting N_2 fixation of peanut (Zuo et al., 2004). Also the nitrate concentration in the soil rhizosphere of intercropped peanut did not increase, nor did the N uptake by peanut compared to sole stand. The authors concluded that the improvement in Fe nutrition was an important factor promoting N_2 fixation by peanut. The competition for N between maize and peanut was not the stimulating factor for N_2 fixation, but the increased Fe availability and uptake by peanut. Both peanut and the root nodule bacteria require Fe for many metabolic functions at several key stages in the symbiotic N_2 fixation. Furthermore, high levels of soil nitrate can be a potent inhibitor of N_2 fixation because then the legumes thrive without fixing atmospheric N (Zuo et al., 2004). Competition for N in a cereal/legume mixture acts as a stimulator for N_2 fixation.

Type II: Single Cropping for Cold Climate and Semi-Arid Crops to Double Cropping for Irrigation Farming

The north-western intercropping region includes the provinces Gansu, Qinghai, Nigxia, Xinjiang, parts of Inner Mongolia and parts of Shaanxi. In comparison to the provinces in the North and the Northeast, the Northwest of China has higher average temperature and more frost-free days, but the average precipitation with 10–250 mm per year is very low. The climate is temperate continental and subtropical, humid in the east and west of the agricultural region changing to dry, continental temperate in the middle. There are long, cold winters and short, hot summers with temperature shifting greatly from day to night as well as from season to season. The crops and varieties grown are fast-maturing ones. Usually, the region was a one harvest-one year district and intercropping wheat with soybean, wheat with millet and maize with faba bean was the main planting system before 1960. From the 1960s onwards, relay intercropping systems with wheat and maize established more and more due to irrigation and varieties improvement. The improvements also led to a gradual shift to a double cropping system in irrigation farming. Although the Northwest is a typical intercropping region with a long intercropping tradition and with intercropping systems being more differential than in the Northeast and the North, the potential for intercropping changes in direction from strip intercropping to relay intercropping and finally to double cropping. Nowadays, intercropping practice in this region is almost all relay intercropping of wheat and maize or wheat and cotton.

Intercropping Wheat or Maize with Legumes
Intercropping wheat with faba bean or soybean increases the yield of wheat in nearly all studies: yield and nutrient acquisition by intercropped wheat and soybean were all significantly greater than for sole wheat and soybean (Li et al., 2001a). Here, intercropping advantages in yield were between 28% and 30% for wheat/soybean. Wheat/soybean had also significant yield increase of intercropped wheat over sole wheat in the study of Zhang and Li (2003). Intercropping resulted in a yield increase in wheat as well as in faba bean between 7% and 46% (Song et al., 2007).

One reason is the effect intercropping has upon N uptake and availability. Zhang and Li (2003) showed that yield increased of about 53% in wheat/soybean, where aboveground effects contributed 23% and belowground effects contributed 30%. For increased N uptake they measured a contribution of 23% aboveground effects and 19% of belowground effects, resulting in an increased overall N uptake of 42%. In contrast to soybean wheat had a greater capability of acquiring nutrients because of the enhanced aggressivity of wheat over soybean. A greater competitive ability and aggressivity of wheat as well as the better nutrient CR led to a greater capability of wheat to acquire nutrients, concerning not only N, but also P and K (Li et al., 2001a). The N accumulation by wheat was mainly due to increased border row N uptake which shows that intercropping is based upon an edge effect.

Besides nutrient acquisition, additional components like border row effects contribute to the overyielding of intercropped wheat. Yields of wheat in border rows significantly increased compared with yield in the inner rows or in rows of sole. Zhang and Li (2003) pointed out that out of a 64% overall increase in yield in intercropped wheat, about 33% came from inner-row effects and about 67% came from the border-row effects. The higher crop overyields due to extra sunlight that taller crops receive on their borders. But accordingly, the N and P accumulation in the border row were significantly greater than in inner row or in sole wheat. Both, border row and inner row contributed to the increase in yield. Studies of Cruse (1996), Ghaffarzadeh (1999), Leopold Center (1995) and Zhang and Li (2003) mentioned that four to six rows seem to be the optimum. Up to six rows a mixed stand is comparable to a sole stand.

The main advantage of intercropping wheat with a legume like soybean or faba bean is the complementary N use; that means, wheat competes much better for soil available N than the legume and, conversely, the legume is forced to get nitrogen from atmospheric N fixation. The competition from wheat in acquiring N through intermingling roots enhanced N_2 fixation in faba bean by about 90% (Xiao et al., 2004). In addition, there is a small N transfer from faba bean to wheat between 1.2% and 5.1% of faba bean N, according to the used measuring techniques, the root distance and contact. This supports the hypotheses of N-sparing by faba bean due to increased N_2 fixation and the increased resource-use efficiency by cropping wheat and a legume together.

Nevertheless, the enhanced N availability or uptake of a cereal crop combined with a leguminous plant like faba bean is more investigated by intercropped maize than intercropped wheat. As shown, the belowground interactions between intercropped species can be more important than aboveground interactions. Intermingling of roots makes sure that nutrients can be used more efficiently because of different mobilizing processes, leading to a higher yield (Li et al., 2007; Zhang and Li, 2003; Zhang et al., 2004). Intermingling of maize and faba bean roots increased N uptake by both crop species by about 20% compared with complete or partial separation of the root system (Li et al., 2003b). The N uptake of faba bean was higher than sole cropped faba bean during early growth stages and at maturity, whereas N uptake of maize did not differ from that by sole maize at maturity, except when P fertilization was high. Because organic acids and protons released by faba bean can mobilize P by the acidification of the rhizosphere, both the N and P uptake by intercropped maize was found to be improved compared with corresponding sole maize (Zhang et al., 2004). Hence, the improved N and P nutrition by intercropping could be characterized as a synergistic process.

A mixture of exudates released from two instead of only one species could change rhizosphere conditions being responsible for the enhanced availability of nutrients, e.g. phosphorus. Zhang and Li (2003) showed in a pot experiment that chickpea facilitated P uptake by associated wheat. Wheat prefers inorganic P and is less able to use organic P. In contrast, chickpea is able to use both P resources effectively. As chickpea mobilizes organic P by releasing phosphatase into the soil turning organic P into inorganic P, P gets available for wheat. Because wheat has a greater competitive ability than chickpea, wheat acquires more inorganic P than chickpea so that chickpea is forced to mobilize organic P. Hence, competition turns into facilitation, because both species do not suffer in P supply. According to the N uptake, P uptake is a combination between above- and belowground effects and interactions. Aboveground interactions contributed 26% to a wheat/soybean mixture, whereas 28% of belowground interactions contributed to the grain yield of wheat (Zhang and Li, 2003).

There are various studies dealing with P supply within intercropping systems, especially maize/chickpea (Inal et al., 2007; Li and Zhang, 2006; Li et al., 2004) and maize/faba bean (Li et al., 2003a/2007; Zhang and Li, 2003). But in most cases, those studies were pot experiments in a greenhouse and not field experiments. Experiments with chickpea promote scientific knowledge of plant nutrition being different whether there are inter- or monocropping conditions, but indeed chickpea is not popular in Chinese cropping systems. Nevertheless, similar to wheat grown together with chickpea, maize intercropped with chickpea too profits by the intermingling of their root systems. Both, faba bean or chickpea and maize accumulated more P in the shoot when intercropped (Li et al., 2003a). The total P uptake by intercropped maize supplied with phytate was twofold greater compared to a monoculture (Li et al., 2004). Li et al. (2004) indicated clearly that the improved growth of maize when intercropped with chickpea was not caused by better N nutrition, but better P uptake. According to Li et al. (2003/2007) four explanations for the increased P uptake by diverse species are plausible:

(1) Greater phosphatase activity in the rhizosphere in intercropping decomposed soil organic P into an inorganic form, which can be used by both species.

(2) Improved P nutrition in maize could have resulted from an increased uptake of P released during the decomposition of root residues after the harvest of e.g. faba bean.

(3) Faba bean, for example, was better nodulated when intercropped, resulting in more fixed N_2. While fixing atmospheric N, legumes take up more cations than anions and release H+ from the roots. Again, H+ is important in dissolving P in calcareous soils.

(4) The volume of soil exploited by the maize roots increased and led to a greater ability to absorb P.

But considering intercropped wheat, Song et al. (2007) went beyond the N accumulation in a plant and studied the community composition of ammonia-oxidizing bacteria in the rhizosphere of wheat and faba bean at different growth stages. Autotrophic ammonia oxidizers in the rhizosphere carried out the first and rate limiting step of nitrification, the oxidation of ammonia to nitrate. The authors concluded that these bacteria could play a key role in N availability to plants and could be important for the interactions between plant species in intercropping. During anthesis the nitrate concentrations in the rhizosphere of wheat intercropped with faba bean were nearly twice as high as in monocropped wheat. Song et al. (2007) suggested that N released from faba bean roots was rapidly mineralized to ammonia and then transformed to nitrate.

Intercropping is known to suppress weeds and pests, because of the higher biodiversity in comparison to monoculture. The soil is covered nearly all the time, and the different plants give home to predators. Most studies dealing with the influence of intercropping on pests and weeds investigated these aspects from the point of view of ecology and less from the point of relationships within a cereal–cereal mixture. Li (2001) reported wheat–cotton relay intercropping being able to control the cotton aphid as well as cotton–rape intercropping that reduced insect damage. The cotton aphid is the main pest of cotton and appears in May. In early May the cotton aphid's natural enemy is the seven-point lady beetle, which is also the natural enemy of the rape aphid. This supports strongly the enemies' hypotheses (Andow, 1991) where the intercropping changes the environmental conditions in such a way that the natural enemy activity is increased. Ma et al. (2007) showed that parasitism of *Allothrombium ovatum* on alate aphids can significantly control the population increase of wheat aphids. Within a strip intercropping of wheat together with alfalfa they examined the possibility to improve the biological control of the wheat aphid by the mite *A. ovatum*. The strip intercropping resulted in higher soil moisture, shadier soil surface and thus a changed microclimate which caused adult female mites to lay more egg pods. In addition, the non-furrow areas provided a more suitable habitat for mites' overwintering, so that the mean number of mites per parasitized aphid was significantly higher in intercrops than in monoculture (Ma et al., 2007).

The different microclimate in intercropping compared to monocropping seems to be substantial for suppressing or enhancing diseases. Chen et al. (2006) showed, that under zero N fertilization, the appearance of powdery mildew in a field was similar in intercropped and monocropped cultivation of wheat and faba bean. They supposed that under deficient N circumstances, plant growth is limited, thus leading to a comparable microclimate within both cropping systems. However, under increased N application rates, the microclimate differs regarding velocity of air movements and correspondingly lower humidity. Chen et al. (2006) concluded that conditions that prevailed in intercropped wheat with faba bean are less conductive to infection by and growth of the powdery mildew compared with sole wheat, because the differences in disease incidence and disease severity due to intercropping between zero and increased N application were significant.

Type III: Double Cropping with Potential for Relay Intercropping
Cereal – especially maize and wheat – production in the Yellow-Huai River Valley is mostly practiced as relay intercropping and less as strip intercropping. Wheat, maize and cotton are the most important and stable crops grown in parts of Hebei, Shanxi, Shaanxi, Anhui and Jiangsu and in Shandong and Henan province. The region is consequently China's granary. With progress in engineering, breeding and irrigation, relay intercropping within the widespread and current double cropping systems is more and more decreasing. In addition, strip intercropping is the only way practised in vegetable production, agroforestry and in fields along big roads. Single cropping with potential for intercropping is common practice in rainfed upland farming, whereas double cropping in large scale or three harvests in 2 years as a more adjusted and sustainable production are to be found in irrigation farming. For the double cropping system, resistant varieties are needed, because of the narrow crop rotation consisting nearly completely of wheat and maize or cotton. Although, summer maize varieties have to be fast maturing because of the very short growing season between June and September/October. Additionally, for implementing two harvests within one year – wheat in June and maize in September/October – irrigation in springtime would be necessary. About 60–70% of the rain falls during the hot summer. In contrast, the winters are cold and dry. Within the intercropping region, the climate changes from temperate continental to subtropical-humid with clear-cut seasons and plum rains between spring and summer.

4. Chapter I

Intercropping Wheat with Maize

Intercropping a cereal–cereal association such as wheat and maize become increasingly popular in irrigated areas and in the North China Plain. Both species grow together for about 70–80 days and yield more than 12,000 kg ha^{-1} (Zhang and Li, 2003). Consequently, both species compete strongly for N and light during their co-growth. Because there is no leguminous plant involved or no facilitation concerning P or Fe supply, the competition seems to be more intrinsic. Nevertheless, scientific research (Li et al., 2001a,b; Zhang and Li, 2003) found that grain yield of both species increased and the N uptake and nutrient accumulation was greater than that by corresponding sole cropping under the same N supply. Indeed, during co-growth biomass yield and nutrient acquisition of the earlier sown wheat increased significantly, whereas at wheat harvest, the biomass of maize in the border row was significantly smaller than in sole maize (Li et al., 2001a). But there is a recovery-compensation growth with the result that at maize maturity the disadvantages disappeared with no significant differences in biomass between border-row maize and inner-row maize or corresponding sole maize. First, the subordinate plant suffers but after harvesting the dominant plant is able to compensate. This competition recovery production principle rests upon the ecological mechanism of niche differentiation. Li et al. (2001b) suggested that interspecific interactions shifted the peak nutrient requirement of dominated species like maize to after wheat harvest, which was helpful for reducing interspecific competition during co-growth. This principle is suitable to intercrop short-season species together with long-season species.

In addition, the more efficient and temporal distribution of soil nutrient consumption results in a reduced nitrate content in the soil profile (Zhang and Li, 2003). The NO_3 – amounts after wheat harvest were greatest under sole wheat and smallest under maize intercropped with wheat (Li et al., 2005). The decrease could be amounting to 30–40% in a wheat/maize association compared to wheat or maize sole (Zhang and Li, 2003). Hence, intercropping can reduce nitrate accumulation and eluviation compared to monocropping.

Like biomass production, grain yield and nutrient accumulation, root growth and root distribution show compatibility of species and niche differentiation of plants. At symmetric interspecific competition, where both species are on an equal footing for acquiring factors of growth, e.g. maize intercropped with faba bean, the spatial root distribution is compatible and similar under inter- as well as monocropped cultivation. Faba bean had a relatively shallow root distribution, maize roots spread underneath them. The shallow root distribution of faba bean results in lower competition for soil resources with the deeper-rooted maize. In contrast, asymmetric competition, where one species dominates over another, e.g. wheat intercropped with maize, results from the greater root proliferation of overyielding species underneath the other. Li et al. (2006) showed that intercropped wheat had a greater root length density compared to sole-cropped wheat, occupied a larger soil volume and extended under maize roots. Roots of intercropped maize were limited laterally to about 20 cm, whereas roots of sole-cropped maize spread laterally about 40 cm. The failure of maize to extend into the soil immediately under wheat may help to explain why maize does not respond positively to intercropping until after the wheat harvest (Li et al., 2006).

Type IV: Three Cropping Seasons per Year with Rotations Replacing Intercropping
Cultivation spectrum and species diversity is much wider in the southwest region, including parts of Guangxi and Shaanxi as well as the provinces Sichuan, Chongqing, Guizhou and Yunnan. Among others, rice, maize, wheat, sweet potato, sorghum, rapeseed, sugarcane, peanut, tea, sesame, bean, vegetable, cotton, ambary hemp, tobacco, millet, cassava, soybean and pea are cultivated, mostly within a rotation and a double or even triple cropping system. In southern parts, double and triple cropping including paddy fields get more and more common. In medium and high plateau uplands, double cropping as well as single cropping for paddy fields are practiced. The production account of wheat decreases more and more when going southward, hence rice and paddy fields increase. Climate conditions allow more flexible rotations, thus replacing and reducing intercropping potentials. Nearly the whole year is frost free with an average temperature between 15°C and 18°C. From the North of the intercropping region to the South, the climate changes from subtropical, humid, monsoonal with intense sunlight and long, hot summers but low temperatures to a more and more tropical climate with conspicuous dry and rainy seasons. Average precipitation is between 800 mm and 1600 mm per year with Guangxi being one of China's most rainy areas. Rainy season is between May and October with occasional droughts in springtime.

CONCLUSION
In conclusion this chapter shows that intercropping of cereals has a 1000-year old tradition in China and it is still widespread in modern Chinese agriculture. As sustainability is the major challenge for Chinese agriculture, intercropping bears the potential for a more sustainable land use without introducing a new cropping system. Nevertheless, there are still research gaps considering intercropping pattern improvements. This chapter is a first approach to take stock of intercropping history, practice and distribution in China. Mostly, these factors are only worth footnotes in international literature. Intercropping regions, area, cropping conditions and crops are rarely part of statistics or reviews. The evaluation of available data on intercropping in this chapter has shown that a classification of four Chinese cereal intercropping regions may be possible, even if it is, in fact, an interim and general classification. More detailed data and studies are needed for further

specification of intercropping regions and patterns. The four intercropping regions are the Northeast and North, the Northwest, the Yellow-Huai River Valley and the Southwest. Going from north to south the cropping systems change from one crop a year with a great potential for intercropping to relay intercropping of especially maize and wheat and double cropping systems and at least three cropping seasons per year with different kinds of rotations and rotations replacing intercropping.

Maize and wheat cultivation is well documented and the main topic of Chinese studies with irrigation and fertilization practice and improvement taking precedence over interspecific competition. Both crops are mostly intercropped with each other or intercropped with legumes and are common all over Chinese intercropping regions. In most studies, maize yields increased when intercropped. Nutrient efficiency increased while N input could be reduced and simultaneously leaching could be reduced because of lower nitrate in the soil. It seems as if maize is a suitable species for intercropping systems in China. Within the double cropping system of wheat and maize, relay intercropping of wheat and maize is common practice. As the region is China's granary the irrigation area and the fertilizer consumption is the highest compared to other intercropping regions, and it is important to reduce input factors and to produce more sustainable output. Improved and adjusted intercropping systems could contribute, e.g. to replace a single maize cycle with an intercropping system, especially in the system of three harvests in 2 years.

The range of species which can be grown in China, especially in the middle and the south of China, is wide, so that there are various possibilities and combinations of intercropping systems. The peanut production in the northeastern region showed as well that intercropping could compensate soil property deficits by reducing Fe chlorosis in peanut plants when intercropped with maize. As intercropping systems offer great opportunities for a more sustainable cultivation, these systems have to be studied more closely in future.

With floods, droughts, landslides and the snow disaster in February 2008, the consequences of environmental pollution will come more and more into mind. In this context, high-input agriculture and monocropping may not be the best performing systems any more, considering income security, nutritional diversity in rural areas, and possibility of severe impacts to large areas due to pest and disease outbreak in a changing climate.

ACKNOWLEDGEMENTS

The authors' research topic is embedded in the International Research Training Group of the University of Hohenheim and China Agricultural University, entitled "Modeling Material Flows and Production Systems for Sustainable Resource Use in the North China Plain". We thank the German Research Foundation (DFG) and the Ministry of Education (MOE) of the People's Republic of China for financial support.

REFERENCES

Andow, D.A. (1991) Vegetational diversity and arthropod population response. Annual Review of Entomology, 36, 561–586.

Atlas of the People's Republic of China (1989) 1st ed., Foreign Languages Press, China Cartographic Pub. House, Beijing.

Bauer, S. (2002) Modeling competition with the field-of-neighbourhood approach – from individual interactions to population dynamics of plants, Diss. an der Philipps-Universität Marburg, Leipzig.

Beets, W.C. (1982) Multiple cropping and tropical farming Systems. Boulder, C.: Westview Pr (u.a.).

Binder, J., Graeff, S., Claupein, W., Liu, M., Dai, M., Wang, P. (2007) An empirical evaluation of yield performance and water saving strategies in a winter wheat–summer maize double cropping system in the North China Plain. Pflanzenbauwissenschaften, 11 (1), 1–11.

Böning-Zilkens, M.J. (2004) Comparative appraisal of different agronomic strategies in a winter wheat–summer maize double cropping system in the North China Plain with regard to their contribution to sustainability. Berichte aus der Agrarwissenschaft D, 100 (Diss. Universität Hohenheim), Aachen.

Chen, Y., Zhang, F., Tang, L., Zheng, Y., Li, Y., Christie, P., Li, L. (2006) Wheat powdery mildew and foliar N concentrations as influenced by N fertilization and belowground interactions with intercropped faba bean. Plant Soil, DOI 10.1007/s11104-006-9161-9.

Chinese Academy of Sciences (1999) (Eds.) The National Physical Atlas of China, China Cartographic Publishing House, Beijing.

Cohen, R. (1988) (Ed.) Satisfying Africa's food needs: food production and commercialization in African agriculture. Book series: Carter studies on Africa, Boulder (u.a.); Rienner.

Cruse, R.M. (1996) Strip Intercropping: A CRP Conversion Option, Conservation Reserve Program: Issues and Options, Iowa State University publication, University Extension, CRP – 17.

Ellis, E.C.,Wang, S.M. (1997) Sustainable Traditional Agriculture in the Tai Lake Region of China. Agriculture, Ecosystems and Environment, 61, 177–193.

FAO (2007) (Eds.) Agricultural Production System Zone Code of the PR. China. Communication Division, Rome.

FAO (2006) (Eds.) The State of Food and Agriculture, FAO Agriculture Series 27, Rome.

Federer, W.T. (1993) Statistical Design and Analysis for Intercropping Experiments, Volume I: Two Crops, Springer Series in Statistics, New York – Berlin – Heidelberg – London – Paris – Tokyo – Hong Kong – Barcelona – Budapest.

Fukai, S., Trenbath, B.R. (1993) Processes determining intercrop productivity and yields of component crops. Field Crops Research, 34, 247–271.

Gale, F. (2002) China at a Glance; A Statistical Overview of China's Food and Agriculture, China's food and Agriculture: Issues for the 21st Century/AIB-775, ed. by Economic Research Service/USDA, pp. 5–9.

Ghaffarzadeh, M. (1999) Strip intercropping, Iowa State University publication, University Extension, Pn 1763.

Gong, Y., Lin, P., Chen, J., Hu, X. (2000) Classical Farming Systems of China. Journal of Crop Production, 3(1), 11–21.

Guohua, X., Peel, L.J. (1991) The agriculture of China, Published in conjunction with the Centre for Agricultural Strategy, University of Reading, New York.

Hauggaard-Nielsen, H., Andersen, M.K., Jørnsgaard, B., Jensen, E.S. (2006) Density and relative frequency effects on competitive interactions and resource use in pea-barley intercrops. Field Crops Research, 95, 256–267.

Inal, A., Gunes, A., Yhang, F., Cakmak, I. (2007) Peanut/maize intercropping induced changes in rhizosphere and nutrient concentrations in shoots. Plant Physiology and Biochemistry, 45, 350–356.

Innis, D.Q. (1997) Intercropping and the Scientific Basis of Traditional Agriculture, Intermediate Technology Publications Ltd., London.

Jolliffe, P.A. (1997) Are mixed populations of plant species more productive than pure stands? Acta Oecologica Scandinavica (OIKOS) 80:3, issued by the Nordic Society, pp. 595–602.

Keating, B.A., Carberry, P.S. (1993) Resource capture and use in intercropping: solar radiation. Field Crops Research, 34, 273–301.

Kiniry, J.R., Williams, J.R., Gassman, P.W., Debaeke, P. (1992) General, process-oriented model for two competing plant species, Transactions of the American Society of Agricultural Engineers (ASAE) 35 (3): general edition; selected contributions in the field of agricultural engineering, pp. 801–810.

Lei C. (2005) Statement, Promoting environmentally-friendly agricultural production in China, Resource Management for Sustainable Intensive Agriculture Systems, International Conference in Beijing, China, April 5–7, 2004, Ecological Book Series – 1, Ecological Research for Sustaining the Environment in China, issued by UNESCO, Beijing, pp. 18–21.

Leopold Center (1995) (Eds.) Potential economic, environmental benefits of narrow strip intercropping. Leopold Center Progress Reports, 4, 14–19.

Li, L., Li, S.M., Sun, J.H., Zhou, L.L., Bao, X.G., Zhang, H.G., Zhang, F. (2007) Diversity enhances agricultural productivity via rhizosphere phosphorus facilitation on phosphorusdeficient soils. PNAS, 104(27), 11192–11196.

Li, L., Sun, J., Zhang, F., Guo, T., Bao, X., Smith, F.A., Smith, S.E. (2006) Root distribution and interactions between intercropped species. Oecologia, 147, 280–290.

Li, L., Sun, J., Zhang, F., Li, X., Yang, S., Rengel, Z. (2001a) Wheat/maize or wheat/soybean strip intercropping I. Yield advantage and interspecific interactions on nutrients. Field Crops Research, 71, 123–137.

Li, L., Sun, J., Zhang, F., Li, X., Rengel, Z., Yang, S. (2001b) Wheat/maize or wheat/soybean strip intercropping II. Recovery or compensation of maize and soybean after wheat harvesting. Field Crops Research, 71, 173–181.

Li, L., Tang, C., Rengel, Z., Zhang, F. (2003a) Chickpea facilitates phosphorus uptake by intercropped wheat from an organic phosphorus source. Plant and Soil, 248, 297–303.

Li, L., Zhang, F. (2006) Physiological Mechanism on Interspecific Facilitation for N, P and Fe Utilization in Intercropping Systems, 18th World Congress of Soil Science, July 9–15, 2006 – Philadelphia, USA, Saturday 15 July 2006, 166–135, papers.

Li, L., Zhang, F., Li, X., Christie, P., Sun, J., Yang, S., Tang, C. (2003b) Interspecific facilitation of nutrient uptake by intercropped maize and faba bean. Nutrient Cycling in Agroecosystems, 65, 61–71.

Li, S.M., Li, L., Zhang, F.S., Tang, C. (2004) Acid Phosphatase Role in Chickpea/Maize Intercropping. In: Annals of Botany, 94, 297–303.

Li, W. (2001) Agro-ecological farming systems in China, Man and the biosphere series, ed. By J.N.R. Jeffers, v. 26, Paris.

Li, W., Li, L., Sun, J., Guo, T., Zhang, F., Bao, X., Peng, A., Tang, C. (2005) Effects of intercropping and nitrogen application on nitrate present in the profile of an Orthic Anthrosol in Northwest China. Agriculture, Ecosystems and Environment, 105, 483–491.

Li, X. (1990) Recent development of land use in China. GeoJournal, 20(4), 353–357.

Lin, J.Y. (1998) How Did China Feed Itself in the Past? How Will China Feed Itself in the Future? Second Distinguished Economist Lecture, Mexico, D.F. CIMMYT.

Lu, C.H., Van Ittersum, M.K., Rabbinge, R. (2003) Quantitative assessment of resource-use efficient cropping systems: a case study for Ansai in the Loess Plateau of China. European Journal of Agronomy, 19, 311–326.

Lu, W., Kersten, L. (2005) Chinese Grain Supply and Demand in 2010: Regional Perspective and Policy Implications. Landbauforschung Völkenrode, 1/2005 (55), 61–68.

Lüth, K.-M., Preusse, T. (2007) China bewegt die Welt. DLG-Mitteilungen, 8, 12–23.

Ma, K.Y., Hao, S.G., Zhao, H.Y., Kang, L. (2007) Strip intercropping wheat and alfalfa to improve the biological control of the wheat aphid *Macrosiphum avenae* by the mite *Allothrombium ovatum*. Agriculture, Ecosystems and Environment, 119, 49–52.

Marschner, H. (1986) Mineral nutrition of higher plants. London [u.a.]; Acad. Pr.

Meng, E.C.H., Hu, R., Shi, X., Zhang, S. (2006) Maize in China; Production Systems, Constrains, and Research Priorities. International Maize and Wheat Improvement Center (CIMMYT), Mexico.

Ministry of Agriculture of the People's Republic of China (2004) (Eds.) Report on the State of China's Food Security, Beijing.

Piepho, H.P. (1995) Implications of a simple competition model for the stability of an intercropping system. Ecological Modelling, 80, 251–256.

Prabhakar, S.V.R.K. (2007) Review of the Implementation Status of the Outcomes of the World Summit on Sustainable Development – an Asia-Pacific Perspective; Climate Change Implications for Sustainable Development: Need for Holistic and Inclusive Policies in Agriculture, Land, Rural Development, Desertification, and Drought, Paper for the Regional Implementation Meeting for Asia and the Pacific for the sixteenth session of the Commission on Sustainable Development,

United Nations Economic and Social Commission for Asia and the Pacific in collaboration with FAO Regional Office for Asia and the Pacific, UNCAPSA, UNCCD Asia Regional Coordinating Unit and UNEP Regional Office for Asia and the Pacific, Jakarta.

Ren, T. (2005) Overview of China's Cropping Systems, Promoting environmentally-friendly agricultural production in China, Resource Management for Sustainable Intensive Agriculture Systems. International Conference in Beijing, China, April 5–7, 2004, Ecological Book Series – 1, Ecological Research for Sustaining the Environment in China, issued by UNESCO, Beijing, 96–101.

Rossiter, D.G., Riha, S.J. (1999) Modeling plant competition with the GAPS object-oriented dynamic simulation model. Agronomy Journal, Reprint April.

Sanders, R. (2000) Prospects for Sustainable Development in the Chinese Countryside; The political economy of Chinese ecological agriculture, Aldershot – Brookfield USA – Singapore – Sydney.

Song, Y.N., Marschner, P., Li, L., Bao, X.G., Sun, J.H., Zhang, F.S. (2007) Community composition of ammonia-oxidizing bacteria in the rhizosphere of intercropped wheat (*Triticum aestivum* L.), maize (*Zea mays* L.) and faba bean (*Vicia faba* L.), Biol Fertil Soils, DOI 10.1007/s00374-007-0205-y.

Su, P., Zhao, A., Du, M. (2004) Functions of different cultivation modes in oasis agriculture on soil wind erosion control and soil moisture conservation. Chinese Journal of Applied Ecology, 15(9), 1536–1540.

Tong, C., Hall, C.A.S., Wang, H. (2003) Land use change in rice, wheat and maize production in China (1961–1998). Agriculture, Ecosystems and Environment, 95, 523–536.

Tsubo, M., Walker, S., Mukhala, E. (2001) Comparisons of radiation use efficiency of mono-/intercropping systems with different row orientations. Field Crops Research, 71, 17–29.

Vandermeer, J. (1989) The Ecology of Intercropping. Cambridge – New York – New Rochelle – Melbourne – Sydney.

Wen, D., Tang, Y., Zheng, X., He, Y. (1992) Sustainable and productive agricultural development in China. Agriculture, Ecosystems and Environment, 39, 55–70.

Wubs, A.M., Bastiaans, L., Bindraban, P.S. (2005) Input Levels and intercropping productivity: exploration by simulation. Plant Research International, note 369 Wageningen, October.

Xiao, Y., Li, L., Zhang, F. (2004) Effect of root contact on interspecific competition and N transfer between wheat and fababean using direct and indirect 15N techniques. Plant and Soil, 262, 45–54.

Xiaofang, L. (1990) Recent development of land use in China. GeoJournal, 20.4, 353–357.

Yamazaki, Y., Kubota, J., Nakawo, M., Mizuyama, T. (2005) Differences in Water Budget among Wheat, Maize and Their Intercropping Field in the Heihe River Basin of Northwest China, AOGS Paper, Asia Oceania Geoscience Society (AOGS) 2nd Annual Meeting 20 to 24 June 2005 in Singapore.

Ye, Y., Li, L., Zhang, F., Sun, J., Liu, S. (2004) Effect of irrigation on soil NO_3 –N accumulation and leaching in maize/barley intercropping field. Transactions of the Chinese Society of Agricultural Engineering, 20(5), 105–109.

Ye, Y., Sun, J., Li, L., Zhang, F. (2005) Effect of wheat/maize intercropping on plant nitrogen uptake and soil nitrate nitrogen concentration. Transactions of the Chinese Society of Agricultural Engineering, 21(11), 33–37.

Yokozawa, M., Hara, T. (1992) A canopy photosynthesis model for the dynamics of size structure and self-thinning in plant populations. Annals of Botany, 70, 305–316.

Yu, C., Li, L. (2007) Amelioration of nitrogen difference method in legume intercropping system. China Scientifical Paper Online, project no.: 20040019035, 1–9.

Zhang, F., Li, L. (2003) Using competitive and facilitative interactions in intercropping systems enhances crop productivity and nutrient-use efficiency. Plant and Soil, 248, 305–312.

Zhang, F., Shen, J., Li, L., Liu, X. (2004) An overview of rhizosphere processes related with plant nutrition in major cropping systems in China. Plant and Soil, 260, 89–99.

Zhen, L., Routray, J.K., Zoebisch, M.A., Chen, G., Xie, G., Cheng, S. (2005) Three dimensions of sustainability of farming practices in the North China Plain; A case study from Ningjin County of Shandong Province, PR China. Agriculture, Ecosystems and Environment 105, 507–522.

Zheng, Y., Zhang, F., Li, L. (2003) Iron availability as affected by soil moisture in intercropped peanut and maize. Journal of Plant Nutrition, 26(12), 2425–2437.

Zuo, Y., Liu, Y., Zhang, F., Christie, P. (2004) A study on the improvement of iron nutrition of peanut intercropping with maize on nitrogen fixation at early stages of growth of peanut on a calcareous soil. Soil Science and Plant Nutrition, 50(7), 1071–1078.

5 Chapter II:
A modeling approach to simulate effects of intercropping and interspecific competition in arable crops

PUBLICATION II:

Knörzer, H., Graeff-Hönninger, S., Müller, B.U., Piepho, H.-P., and Claupein, W. (2010): A modeling approach to simulate effects of intercropping and interspecific competition in arable crops. International Journal of Information Systems and Social Change 1 (4), pp. 44-65.

Whereas chapter I was an overview on intercropping in China as such, the second chapter deals with the modeling and simulation of intercropping and interspecific competition. The chapter addresses the questions of: how was knowledge about competition effects used and transferred to crop growth models and simulations so far? Which models dealt already with intercropping and interspecific competition effects? What single or multiple effects were considered, modeled, and in which way? Within this modeling intercropping thesis at hand, chapter I is a literature review about the status quo of 'intercropping', and chapter II is a literature review on the status quo of 'modeling' of those systems. Both studies generated the basis on which further modeling within that thesis was done and assessed. On the one hand, the review showed what has been done and what has been possible to model or simulated intercropping so far. Yet, it showed its limitations as well as, as a result, a first approach to model intercropping of cereals with DSSAT and with a basically different procedure. Thus, it shows an alternative and for further researches a promising way to handle intercropping within modeling and simulation, wherefore the second part of chapter II could be considered as a theoretical framework or concept and a starting point for further intercropping studies with DSSAT as well as other crop growth models. The theoretical approach of integrating a shading algorithm into the model as well as considering microclimate differences within intercropping systems was shown and evaluated using own field data of a wheat-maize intercropping system.

5. Chapter II

A Modeling Approach to Simulate Effects of Intercropping and Interspecific Competition in Arable Crops

Heike Knörzer, Universität Hohenheim, Germany
Simone Graeff-Hönninger, Universität Hohenheim, Germany
Bettina U. Müller, Universität Hohenheim, Germany
Hans-Peter Piepho, Universität Hohenheim, Germany
Wilhelm Claupein, Universität Hohenheim, Germany

Article from the International Journal of Information Systems and Social Change, 1(4), 44-65, October-December 2010, with permission, copyright © 2010, IGI Global.

ABSTRACT

Interspecific competition between species influences their individual growth and performance. Neighborhood effects become especially important in intercropping systems, and modeling approaches could be a useful tool to simulate plant growth under different environmental conditions to help identify appropriate combinations of different crops while managing competition. This study gives an overview of different competition models and their underlying modeling approaches. To model intercropping in terms of neighbouring effects in the context of field boundary cultivation, a new model approach was developed and integrated into the DSSAT model. The results indicate the possibility of simulating general competition and beneficial effects due to different incoming solar radiation and soil temperature in a winter wheat/maize intercropping system. Considering more than the competition factors is important, that is, sunlight, due to changed solar radiation alone not explaining yield differences in all cases. For example, intercropped maize could compensate low radiation due to its high radiation use efficiency. Wheat benefited from the increased solar radiation, but even more from the increased soil temperature.

Keywords: DSSAT, Field Boundary Cultivation, Intercropping, Interspecific Competition, Maize, Solar Radiation, Wheat

INTRODUCTION

Intercropping, defined as growing two or more crops simultaneously on the same field (Federer, 1993), is widespread all over the world. Especially in smallholder farming like in Africa (e.g., Malawi: 80 – 90% of soybean cultivation), India (17% of arable land) or China (25% of arable land), intercropping is a common cropping system. In times of climate change, rising food prices, shortage of arable land and food in third world countries and countries with a rapidly increasing population, adjusted traditional cropping systems become more and more important. Farmers tend to utilize every square centimetre of available arable land for production and for diversification of their families diet. Besides, there is a so-called unconscious intercropping: Because fields and farm size are very small (0.1 – 2 ha), the sum of field borders can be considered as intercropping in a larger scale (Figure 1).

Competition results not only in a survival of the fittest, but also in an optimal use of ecological niches. Agriculture can utilize interspecific competition in order to adjust cropping systems. Some attempts have been made to investigate and improve the various forms of intercropping. An increasing number of these research efforts, especially during the 1990's, were done by modeling studies in order to simulate interspecific competition. Most models dealing with interspecific competition are common crop grow or crop/weed models extended with a submodel or additional algorithms. In most cases, modeling a cereal-cereal interaction, the crops of choice are a cereal-legume mixture as on one hand, this crop combination is a common and widespread intercropping system due to the advantages of nitrogen supply by the legume and on the other hand, these species are already included in most crop growth models.

Nevertheless, intercropping has always been considered as a secluded cropping system within one field so far. But in African and Asian countries, where intercropping is widespread, the system can be extended to a much larger scale: common on-field intercropping goes along with small field size on average, low mechanization level and hence, small field boundary distances. For example in China, where the average farm size is around 0.1 ha, small fields alternate as stripes with different crops grown on it and turning field boundaries into a kind of unconscious intercropping at a larger scale. To simulate not the secluded intercropping system explicitly, but the field boundaries could turn modeling of a single cropping system into modeling of more regional considered cropping patterns.

Competition for light seems to be the most palpable, both, for measurements in the field and for submodeling. However, intercropping cannot be considered solely as a change in available solar radiation within a dominant and understorey canopy, as it influences also soil properties like temperature and moisture, root distribution, microclimate conditions like wind speed and humidity, pests and diseases and nutrient availability for the plants standing next to each other. The possibilities to model intercropping are various: modified weather and climate-, soil-, growth factors and plant health-indices are imaginable. But as data collection in the field is difficult in intercropping systems, modelers often restrict modeling of intercropping to competition for solar radiation (Ball and Shaffer, 1993; Baumann et al., 2002; Lowenberg-DeBoer et al., 1991; Wiles and Wilkerson, 1991). Even there, weekly plant samples and data collection or samples in more frequent intervals are necessary. However, most models simulate the effect of intercropping using a similar model approach for simulating e.g., competition for solar radiation (see subsection 'background').

A  B

Figure 1: In China, the average field size is very small and fields alternate as stripes with different crops grown on it, turning field boundaries into a kind of unconscious intercropping at a larger scale (A, B). For illustration, field boundaries are marked with white lines (A) and field length and width are between 5 to 20 m

The development of a general competition algorithm to be introduced as a submodel in existing crop models might be a chance to promote the intercropping research turning from evaluating and validating to adjusting cropping systems or to develop appropriate and improved intercropping systems. Nevertheless, research still strives for finding such a general algorithm. In addition, introducing a generalized submodel is not easy to handle all over the various models and needs sometimes a reprogramming of the model. After all, it would be a competition submodel for solar radiation and not for interspecific competition at all.

Questions about the hitherto existing status quo of modeling interspecific competition have still remained open: which data input is necessary? Which species are modeled so far? And which equations and algorithms are typical and often used for modeling competition? Besides, how comprehensive should those models be either to take different competition factors into account or to be easy in handling? This article gives an overview over existing interspecific competition models and their model behaviour, and can also be considered as a starting point for further modeling work. So far, the existing intercropping models are only at their very beginning and lead the way to more intensive and practice-related studies. The various approaches seem to be promising and complementing each other, but there is still a gap between the modeling of case studies and the application of those models and the adjustment of existing cropping systems, especially to extend them to the aspect of field boundary cultivation.

This paper reviews case studies in which various existing intercropping models have been successfully validated. Around 20 different models are considered for modeling interspecific competition in different ways, for example: ALMANAC, APSIM, ERIN, FASSET, GAPS,

GROWIT, INTERCOM, KMS, NTRM-MSC, SIRASCA, SODCOM, SOYWEED, STICS, VCROPS and WATERCOMP. Applications range from European organic farming systems to the simulation of maize and legumes growth and development in Africa as well as the prediction of performance of intercropped vegetables (Table 1). Based on the evaluation of existing models, research gaps were identified and a model modification for the simulation of intercropping/field boundary cultivation was developed and introduced in the process-oriented crop growth model DSSAT 4.5 (Decision Support System for Agrotechnology Transfer) (Jones et al., 2003), to extent the view from competition parameters like solar radiation to a whole-species view, e.g., to subdivide not only the canopy layers but also the within species effects and to model one species as the sum of subspecies behaviour under different intercropping related environmental circumstances. Our modified model is more generalized, and offers therefore a more comprehensive and integrative approach.

BACKGROUND

There are different strategies introducing interspecific competition or intercropping into existing crop models:

- To supplement, include or link an additional submodel into existing crop growth models, e.g.,
    STICS (Brisson et al., 2004),
    INTERCOM (Baumann et al., 2002; Kropff and van Laar, 1993),
    FASSET (Berntsen et al., 2004),
    ALMANAC (Kiniry et al., 1992).
- To incorporate modifications which take account of the competition between intercrops, e.g.,
    APSIM (Carberry et al., 1996; Nelson et al., 1998),
    SODCOM (O'Callaghan et al., 1994).
- To evaluate new model approaches based on already existing models, e.g.,
    GAPS (Rossiter and Riha, 1999) based on ALMANAC,
    LTCOMP (Wiles and Wilkerson, 1991) based on SOYWEED respectively SOYGRO,
    NTRM-MSC (Ball and Shaffer, 1993) based on NTRM,
    AUSIM (Adiku et al., 1995) based on APSIM.
- To evaluate and validate models which allows the assessment of specific management influences like intercropping, e.g.,
    SUCROS-cotton (Zhang, 2007).
- To extend simple equation models, e.g.,

KMS (Sinoquet et al., 2000) is a simplified Kubelka-Munk equations model which can be extended to multispecies canopies,

VCROPS (Garcia-Barrios et al., 2001) is a further development of Vandermeer's spatially explicit individual-based mixed crop growth model (Vandermeer, 1989),

WATERCOMP is a modified form of the Penman-Monteith equation (Ozier-Lafontaine et al., 1998).

Furthermore, there are a few unnamed models concerning competition among different species. They mainly deal with a single phenomenon of competition rather than a whole field system, e.g.,

- Kropff and Spitters (1992),
- Sellami and Sifaoui (1999),
- Tsubo and Walker (2002),
- Tsubo et al. (2005),
- Yokozawa and Hara (1992).

In addition:

- GROWIT (Lowenberg-DeBoer et al., 1991) is some kind of exception, because there are no sub-model modules used, but instead a spreadsheet template for making stochastic dominance comparisons.
- The model EcoSys (Caldwell, 1995; Grant, 1992, 1994) is mainly used for simulating multispecies ecosystems across a landscape and to represent fundamental physical and physiological processes at the scales used especially for plant physiologists and less for agriculture and crop growth purposes.

Models dealing with interspecific competition get more and more important as the postulation for sustainable agricultural production has become a global political issue. Traditional farming systems like intercropping or mixed cropping are known to be the embryonic form of sustainable production concerning biodiversity, resource use efficiency and yield stability. As field trials are time consuming and expensive, models are the alternatives. They help decision makers by reducing time and human resources as well as researchers to provide a framework for scientific cooperation (Jones et al., 2003).

Intercropping or competition models can be roughly classified: a) those which are able to model plant growth within a field or even farm scale, taking soil, atmosphere and management information into account in a more dynamic and mechanistic way, and b) those which are more specified or simplified to single intercropping phenomenon like partitioning of solar radiation (Tsubo and

Walker, 2002) or self thinning (Yokozawa and Hara, 1992) and are more static and empirical. There are examples for both categories in the literature (Table 1):

a) ALMANAC
   APSIM
   AUSIM
   FASSET
   INTERCOM
   STICS

b) Canopy photosynthesis model in plant populations
   Radiation transmission model and simple model for intercropping

Table 1: Overview over plant and crop growth models simulating interspecific competition and/ or intercropping systems

(- continued on following page -)

5. Chapter II

Model	Cropping System	Model type	Source
ALMANAC	Maize and soybean	Dynamic, process-oriented plant growth, water balance and nutrient balances model	Kiniry and Williams, 1995
APSIM	Maize with leguminous shrub hedgerows for tropical farming systems in developing countries Rotation of sorghum/maize or crops/ley pasture or crop with an understorey of volunteer legume in northern Australia	Flexible software system for simulating agricultural production systems, different soil, biological and managerial processes are taken into account arising from interactions between different crops and pasture grown in rotation	Carberry et al., 1996; Nelson et al., 1998
AUSIM	Maize and cowpea	Morphological, physiological and phonological model by linking the respective sole crop models	Adiku et al., 1995
FASSET	Pea and spring barley	Dynamic whole-farm model	Berntsen et al., 2004
GAPS	Dynamic simulation of inter-species competition in agricultural systems	Dynamic model of the soil-plant-atmosphere systems where multiple plant species are grown in competition	Rossiter and Riha, 1999
GROWIT	Millet-cowpea	Generic framework using a spreadsheet for making stochastic dominance comparisons, estimates plant growth by integrating over a continuous growth function depending on air temperature	Lowenberg-De-Boer et al., 1991
INTERCOM	Celery and leek	Process-based eco-physiological model, simulates dynamically competition processes based on physiological, morphological and phenological processes	Baumann et al., 2002
SODCOM	Maize and beans in Kenya	Dynamic and mechanistic model taking especially physiological, morphological and phenological as well as soil moisture components into account	O'Callaghan et al., 1994
STICS	Pea and barley in European organic farming *Gliricidia sepium*, natural pasture C_4 grass and maize/canavalia and maize/sorghum	Whole-field soil-plant atmosphere model over one or several crop cycles, simulates a crop situation for which a physical medium and a crop management schedule can be determined	Brisson et al., 2004; Jensen, 2006
VCROPS	Greenhouse diculture with radish and bushbean	A further developed spatially explicit individual-based mixed crop growth model to simulate individual plant growth and to perform statistical analysis of deterministic and stochastic versions of the model	García-Barrios et al., 2001

continued on following page

Table 1. continued

WATERCOMP	Maize and sorghum in Central-America	Physically based framework including a radiative transfer model associated with a transpiration-partitioning model with special regard to spatial aspects of root competition at the scale of static root systems	Ozier-Lafontaine et al., 1998
Canopy photosynthesis model in plant populations	Not specified	Dynamic model for growth and mortality of individual plants in a stand assuming an even-aged plant population which grows in a homogenous environment	Yokozawa and Hara, 1992
Extended IBSNAT models	Beangrow and Ceres-maize models	Hypothetical crop model of simulating a cereal-legume intercrop dealing with only part of the intercropping phenomenon	Thornton et al., 1990
Model for plant and crop growth allowing for competition for light	Cabbage and carrot	Plant growth model applicable to the isolated plant or to even-aged plants, simple mathematical relation to describe the efficiency of light interception	Aikman and Benjamin, 1994
Radiation transmission model and simple model for intercropping	Maize and bean under semi-arid conditions	Instantaneous, radiation transmission model comparison between a geometrical versus a statistical method, simple model to be employed to develop other cereal-legume intercrop models for semi-arid regions	Tsubo and Walker, 2002; Tsubo et al., 2005

Competition between (intercropped) species is mainly competition for water, nutrients and light respectively solar radiation. Furthermore, one can distinguish between above- and below-ground interaction and competition, but also beneficial and synergistic effects can occur (Inal et al., 2007; Li et al., 2001ab; Song et al., 2007; Zhang and Li, 2003). But it is striking that almost all models deal solely with competition for light (Baumann et al., 2002; Brisson et al., 2004; Sellami and Sifaoui, 1999; Tsubo and Walker, 2002; Wiles and Wilkerson, 1991), sometimes in combination with competition for water (Lowenberg-DeBoer et al., 1991; O'Callaghan et al., 1994; Ozier-Lafontaine et al., 1998). Concerning multi-species water and nutrient supply most models use similar or even the same basic and common soil properties, which they use for sole crop situations neglecting aspects and characteristics of competing root systems making some species more efficient for water uptake than others. Beneficial interactions especially for nutrient supply like nitrogen (N), phosphorus (P), potassium (K) and iron (Fe) are rarely taken into account. It is known, however, that the structure and distribution characteristics of roots are more important than their own absorption capacity in the competition for water and solutes (Ozier-Lafontaine et al., 1995).

Light partitioning is the most frequently modeled competition resource. Morphological, physiological and also phenological differences between species decide whether a species has an advantage or a disadvantage in intercropping. In most cases, the earlier developing or the taller species becomes to be the dominant one, the other the understorey. Parts of those understorey species are shaded from the dominant one. The shading impact depends on the species height, their development stage and aggressivity, the duration of competition and the distance from the understorey to the dominant species (row effect). The amount of captured solar radiation decreases for the understorey species and has to be taken into account during modeling. Determinants of light partitioning are vertical dominance and differences in foliage inclination, leaf area and plant height. As solar radiation is reduced, also temperature (Zhang, 2007) within the microclimate of the shaded understorey species can decline as a secondary effect of competition for light.

Light capture ability as well as space occupation influence competition impact and crop performance. Accordingly, radiation models for multispecies canopies can be classified into four groups (Sinoquet and Caldwell, 1995):

1. Geometrical models which schematized plants or rows as simple shapes arranged in space according to the planting pattern.

2. Models based on the turbid layer medium analogy where the canopy structure is described by statistical distributions.

3. Hybrid models combine geometrical shapes as subcanopies envelops with statistical description of leaf area distribution within those envelops.

4. 3D plant descriptions.

For process-oriented models, the turbid layer medium analogy has proven to be the most useful. Almost all simulation models devoted to light partitioning between species are based on the classical Beer's law – but modified for a two-species-purpose - and compute light transmission as a negative exponential function of the downward cumulated leaf area index (LAI) (Table 1).

Table 2: LER[1], RYT[2], RLO[3] and A_{ab}[4] for inter- and monocropped wheat and maize in 2007/08

	Tinsley, 2004 LER[1]	Jolliffe, 1997 RYT[2]	Jolliffe, 1997 RLO[3]	Li et al., 2001a A_{mw}[4]
borderline	1.30	1.16	1.16	-0.819
3./4. row rp. 2-4 m	0.94	0.97	0.97	
5./6. row rp. 2-4 m	0.94	0.97	0.97	
monocropping	1.00	1.00	1.00	

[1] Land equivalent ratio (LER) (Tinsley, 2004) = $[(Y_{1m}-Y_{1i})+(Y_{2m}-Y_{2i})]/100$
  Y_m = yield in monoculture
  Y_i = yield in intercropping
  1, 2 = first and second crop

[2] Relative yield ratio (RYT) (Jolliffe, 1997) = $[(Y_i)_m / (Y_i)_p] + [(Y_j)_m / (Y_j)_m]$
  Y = yield
  i, j = species 1 and 2
  m = species mixture
  p = pure stand

[3] Relative land output (RLO) (Jolliffe, 1997) = $(Y_i + Y_j + \ldots)_m / (Y_i + Y_j + \ldots)_p$

[4] Aggressivity factor (A_{ab}) (Li et al., 2001a) = $[Y_{ia} / (Y_{sa} * F_a)] - [Y_{ib} / (Y_{sb} * F_b)]$
  Y = yield
  s = sole cropping
  i = intercropping
  F = proportion of the area occupied by the crops in intercropping
  A, B = crop 1 and 2

Table 3: Results of the statistical analysis to detect differences between rows/subplots with different distances from the field border or differences between monocropped and intercropped plots indicated that intercropping was a field border effect. Rows/subplots sharing no letter are significantly different at $\alpha = 5\%$

Intercropping system	Grain yield differences between different rows/subplots[1] (t ha^{-1})	
Maize (intercropped with wheat)	Row 1+2^a to row 3+4^a	0.1
	Row 1+2^a to row 5+6^a	0.2
	Row 1+2^a to row 7+8^a	0.4
	Row 3+4^a to row 5+6^a	0.0
	Row 3+4^a to row 7+8^a	0.5
	Row 5+6^a to row 7+8^a	0.5
Maize (intercropped with pea)	Row 1+2^a to row 3+4$^{b, c, d}$	1.5
	Row 1+2^a to row 5+6$^{b, c}$	2.0
	Row 1+2^a to row 7+8^d	1.2
	Row 3+4$^{b, c, d}$ to row 5+6$^{b, c}$	0.5
	Row 3+4$^{b, c, d}$ to row 7+8^d	0.3
	Row 5+6$^{b, c}$ to row 7+8^d	0.8
Maize (intercropped with peanut)	Row 1^a to row 2$^{a, b}$	0.9
	Row 1^a to row 3$^{a, b, c}$	1.4
	Row 1^a to row 4$^{b, c}$	2.6
	Row 1^a to row 5^c	2.7
	Row 1^a to row 6$^{b, c}$	2.5
	Row 1^a to row 7$^{b, c}$	2.2
	Row 2$^{a, b}$ to row 3$^{a, b, c}$	0.5
	Row 2$^{a, b}$ to row 4$^{b, c}$	1.6
	Row 2$^{a, b}$ to row 5^c	1.7
	Row 2$^{a, b}$ to row 6$^{b, c}$	1.6
	Row 2$^{a, b}$ to row 7$^{b, c}$	1.3
	Row 3$^{a, b, c}$ to row 4$^{b, c}$	1.1
	Row 3$^{a, b, c}$ to row 5^c	1.2
	Row 3$^{a, b, c}$ to row 6$^{b, c}$	1.1
	Row 3$^{a, b, c}$ to row 7$^{b, c}$	0.8
	Row 4$^{b, c}$ to row 5^c	0.1
	Row 4$^{b, c}$ to row 6$^{b, c}$	0.0
	Row 4$^{b, c}$ to row 7$^{b, c}$	0.3
	Row 5^c to row 6$^{b, c}$	0.2
	Row 5^c to row 7$^{b, c}$	0.4
	Row Row 6^c to row 7^{b,c}	0.3
Wheat (intercropped with maize)	0-2 m^a to 2-4 m^b	3.4
	0-2 m^a to 4-6 m^b	3.3
	2-4 m^b to 4-6 m^b	0.1
Pea (intercropped with maize)	0-2 m^a to 2-4 m^a	0.05
	0-2 m^a to 4-6 m^a	0.06
	2-4 m^a to 4-6 m^a	0.01

continued on following page

Table 3. continued

Peanut (intercropped with maize)	Row 1[c, d] to row 2[b, c, d]	0.33
	Row 1[c, d] to row 3[c]	0.10
	Row 1[c, d] to rows 4+5[a, d]	0.67
	Row 1[c, d] to rows 6+7[a, b]	0.81
	Row 1[c, d] to rows 8+9[a]	1.22
	Row 2[b, c, d] to row 3[c]	0.42
	Row 2[b, c, d] to rows 4+5[a, d]	0.34
	Row 2[b, c, d] to rows 6+7[a, b]	0.48
	Row 2[b, c, d] to rows 8+9[a]	0.90
	Row 3[c] to rows 4+5[a, d]	0.76
	Row 3[c] to rows 6+7[a, b]	0.91
	Row 3[c] to rows 8+9[a]	1.32
	Rows 4+5[a, d] to rows 6+7[a, b]	0.15
	Rows 4+5[a, d] to rows 8+9[a]	0.56
	Rows 6+7[a, b] to rows 8+9[a]	0.41

NEW APPROACH FOR MODELING INTERCROPPING WITH SPECIAL REGARD TO FIELD BOUNDARY CULTIVATION

In the current literature, intercropping was considered either as the performance of a secluded cropping system within one field or area and with a specific species combination like maize/legumes or as a spatially explicit occurrence of interspecific competition between different species, especially crops/weeds. Driving forces for modeling were the geometrical shape or the influence of vertical distribution of leaves of a species (Baumann et al., 2002; Berntsen et al., 2004; Wiles and Wilkerson, 1991), the assumption of a reduced leaf area index of the understorey species when shading of the dominant species occurs (O'Callaghan et al., 1994; Rossiter and Riha, 1999), the dividing of the canopies into three different canopy layers (Brisson et al., 2004; Jensen, 2006; Sinoquet et al., 1990) and to calculate – or better estimate – the incoming solar radiation for both species according to a modified Beer's law, taking crop coefficients, leaf area and the height of the neighboring plant into account.

The approach in this study refers in a first step (scenario 1) to the traditional approach of using competition for solar radiation as baseline model. In a second step (scenario 2 and 3) not only competition for solar radiation was taken into account, but also a more general approach to make it possible to transfer the basic assumptions from one intercropping system to an overall intercropping system by evaluating a minimum data set, easily being collected in a field and not restricted to one area. Special regard was taken to the so called unconscious intercropping or field boundary cultivation, because of its importance in African or Asian countries where intercropping plays a major role and field sizes are very small. Three different scenarios were studied to evaluate the model. Scenario 1 takes only competition for solar radiation into account, scenario 2 deals with changes in soil albedo and scenario 3 includes besides competition for solar radiation also changes in soil temperature.

Field trials (for detailed description of field trials: see next subsection), conducted in southwest Germany and in the North China Plain with different species combination, showed that if synergistic or competition effects of intercropping occur, they are mainly based on border row effects (Table 3). Significant differences in grain yield, dry matter accumulation, leaf area etc. from monocropped in comparison to intercropped species were mostly restricted to the first four rows wherefore intercropping could be regarded as borderline or field boundary effect.

Table 4: Genetic parameters of wheat and maize and their values used in the model evaluation

winter wheat		
parameters	description	value
P1V	sensitivity to vernalisation	20
P1D	sensitivity to photoperiod	30
P5	grain filling duration	750
G1	kernel number per unit weight at anthesis	25
G2	kernel weight under optimum conditions	40
G3	standard stem and spike dry weight at maturity	0.5
PHINT	phylochron interval	80

maize		
parameters	description	value
P1	growing degree days from emergence to end of juvenile phase	110
P2	photoperiod sensitivity	0.05
P5	cumulative growing degree days from silking to maturity	615
G2	potential kernel number	615.0
G3	potential kernel growth rate	8.2
PHINT	phylochron interval	38.9

MATERIAL AND METHODS

Field trials were conducted in southwest Germany (48.46°N and 8.56°E) at the University of Hohenheim's experimental station 'Ihinger Hof' during 2007/08. The average rainfall per year is around 690 mm with an average temperature of 7.9°C. The soils are mainly keuper with loess layers. Alternate plots of winter wheat and maize were arranged within a restricted randomized complete block design and four replications. Randomization was restricted due to the strip intercropping character of the experiment. Each plot was 10 x 10 m for wheat and 12 x 10 m for maize respectively and included five subplots (2 x 10 m) for wheat and eight subplots (2 rows x 10 m) for maize. The previous crop for all species was sugar beet and the soil preparation was a reduced tillage system with a chisel plough. Within each subplot data was collected in order to detect differences between the boundaries, different distances from the boundary and the monocropping. The plots were big enough to reflect monocropping within the central subplots. Wheat was sown in October 2007 with a row spacing of 13 cm, a row orientation from north to south and a plant density of 300 plants per m^2. Maize was sown in May 2008 with a row spacing of 75 cm, a row orientation from north to south and a plant density of 10 plants per m^2. During the growing season, neither water nor nitrogen stress occurred so that differences in plant growth and yield performance could be attributed onto intra- and interspecific competition. Wheat was fertilized with 160 kg N ha^{-1}, splitted into three dispensations (60/60/40) of Nitro-chalk. Maize was fertilized once with 160 kg N ha^{-1} (ENTEC). Plant protection was carried out according to 'Good Agricultural Practice'.

During the growing season, three temporal harvests were carried out and dry matter, ratio of stems to leaves and the nitrogen concentration of plants were analysed. Between flowering and maturity, three time harvest were taken to determine grain weight growing rates. In addition, N_{min} content of soil and number of plants per m^2 were determined. LAI was measured destructively according to the DSSAT guide with the LI-3100 Area Meter (LI-COR, Lincoln/ Nebraska USA); soil moisture was measured with the Trime-TDR-system (time domain reflectometry) from IMKO GmbH (Ettlingen/ Germany) and soil temperature was determined in 2 and 15 cm soil depth with the testo 925 (Testo AG, Lenzkirch/Germany) on a weekly basis as well as solar radiation, growing stages and plant height. Specific DSSAT cultivar coefficients like the phylochron interval in thermal days, grain filling duration in thermal days, the standard kernel size related to kernel filling rate and the standard dry-weight of a single tiller at maturity for wheat were determined and yield and yield components were measured after the final harvest. Respectively, phylochron interval, kernel filling rate, maximum possible number of kernels per plant and thermal time from seedling emergence to the end of the juvenile phase expressed in degree days were determined for maize (DSSAT 4.5). Furthermore, the land equivalent ratio (LER) and an aggressivity factor (A_{ab}) (Table 2) were calculated for both the intercropping/field boundary and the monocropping. When Aab > 0, the competitive ability of crop A exceeds that of crop B. Concerning maize and wheat, the agressivity factor (A_{mw}) was -0.818 indicating that wheat was the more aggressive competitor and was more successful in catching growing factors, e.g., nitrogen, thus being important for further modeling approaches.

Statistical analysis (Table 3) to detect significance between rows/subplots within a plot with different distances from the plot border was done separately for each species and intercropping system in the trial. Analysis was done using the mixed procedure of SAS 9.2 (SAS Institute, 2009) according the model: trait = replicate + position, assuming independent errors of rows within plots.

The factor position had levels depending on treatment (Table 3). Differences of least squares means for positions were subjected to t-tests. Significant differences found were restricted to the first few rows reflecting the intercropping situation and giving evidence of intercropping being a field border effect (Table 3).

Model Evaluation

For the model approach, the DSSAT crop growth model was applied. It is a process-oriented crop model taking soil-plant-atmosphere and -management systems into account. It is designed for helping researchers to adapt and test the cropping system model itself as well as for those operating the DSSAT model to simulate production over time and space for different purposes (Jones et al., 2003). The collected dataset in this case study was used to evaluate a model extension for intercropping including crop management like fertilization, influence of previous crop (sugar beet), soil and genotype characteristics and weather data (© Hohenheimer Klimadaten). Phenology and growth data were used to evaluate the genetic parameters in a first step (Table 4).

Whereas the genotype and soil characteristics as well as the management did not differ between monocropped and intercropped systems and were taken as constants during simulation, the microclimate changed. Especially the influence of shading and the soil temperature (Figure 2) within the boundaries were different between monocropped and intercropped situations and thus being the basic for the model approach. The approach was carried out in a stepwise fashion, creating different scenarios. Thus, in a first model run, the identical DSSAT project-oriented programming was used for the intercropping model as well as for the monocropping model except for modified solar radiation. In scenario 2, soil albedo was modified in order to take increased soil temperature into account. In scenario 3, top soil temperature in addition with a higher agressivity factor for wheat was taken furthermore into account including the modification of the initial nitrogen conditions.

The increased ability of intercropped wheat (A_{mw}) to acquire more nitrogen (N) in association with the increased soil temperature coming along with an increased mineralization had to be factored in the modeling. At the field boundary, from sowing date until the sowing of maize in May, wheat faced no interspecific competition and less intraspecific competition. Because of the increased top soil temperature within the first rows – 4°C difference on average - and the higher agressivity of wheat in comparison to maize, our hypothesis was that the mineralization of nitrogen might be favoured and intercropped wheat might get more nitrogen than its monocropped equivalent. The previous crop, sugar beet, had an average N surplus of about 55 kg N ha^{-1} (Reisch and Knecht, 1995) after harvest. As the sugar beet leaves were not removed, they could also be calculated as organic amendment comparable to 140 kg N ha^{-1} (Stammdatenblätter "Nährstoffvergleich Feld-Stall", LEL Schwäbisch-Gmünd et al., 2007).

Figure 2: Differences in average soil temperature under monocropped and intercropped wheat in June and July 2008

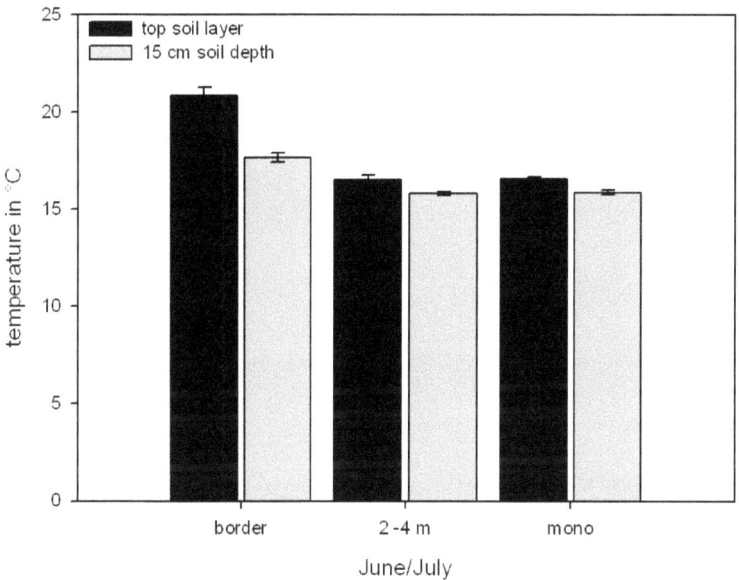

The management system was a reduced tillage system. Lower soil temperature and a reduced mineralization at the beginning of the growing season in comparison to tillage systems are typical for reduced tillage systems (Koeller and Linke, 2001). Collected N$_{min}$-samples after the harvest of wheat showed a similar amount of nitrate in the soil for intercropping (20 kg NO$_3$ ha^{-1}) as well as monocropping (22 kg NO$_3$ ha^{-1}) with only 2 kg NO$_3$ ha^{-1} differences. However, making up the balance between intercropped and monocropped plots concerning nitrogen proportion and taking N$_{min}$ at the beginning of the growing season, N$_{min}$ after harvest, amount of fertilization and N withdrawal from grain and straw into account, there was a N gap between intercropping and monocropping. Wheat border rows got approximately 130 kg N ha^{-1} more until anthesis. Hence, the changed microclimate within the field boundary might lead to a better N supply of wheat in the first rows (Figure 2).

In comparison to most other modeling approaches, changes in incoming solar radiation were neither calculated according to a modified Beer's law nor by dividing the canopy into different layers nor assuming that LAI was reduced from the time shading occurs. Instead, a linear shading pattern in percentage based on weekly sunlight measurements was determined with regard to the height of neighbouring plants. Solar radiation of monocropped species was set as 0% shading meaning that the monocropping system was driven by the original weather data. Then, the light differences of the border rows in proportion to the monocrops were calculated. The shading pattern was subjected to the plant height of the neighbouring species resulting in a linear function given as:

- Shading pattern within wheat: (1a)

  $s = 0.2278h + 33.417$ ($R^2 = 0.6708$),

  with s = shading of wheat border rows (%),

  h = plant height of maize.

- Shading pattern within maize: (1b)

  $s = -19.724h + 1594.2$ ($R^2 = 0.975$),

  with s = shading of maize border rows (%),

  h = plant height of wheat.

Logistic models were fitted by nonlinear least squares using the NLIN procedure of the SAS System:

- Plant height of wheat: (2a)

  $h = 85.9882/(1 + \exp(4.7888 - 0.059 * DAS))$,

  with h = plant height,

  DAS = days after sowing.

- Plant height of maize: (2b)

  $h = 300.5/(1 + \exp(13.1402 - 0.073 * DAS))$,

  with h = plant height,

  DAS = days after sowing.

The daily solar radiation (SRAD) expressed in $MJ/m^2/d$ was calculated according to (1) and (2) and modified for the weather input as follows:

(3a)

$SRAD_{inter/wheat} = SRAD_{mono} - ((0.2278 * (300.5/(1 + EXP(4.7888 + (-0.073) * DAS))) - 33.417) * SRAD_{mono}/100)$

(3b)

$SRAD_{inter/maize} = SRAD_{mono} - ((-19.724 * (85.9882/(1 + EXP(13.1402 + (-0.059) * DAS))) + 1594.2) * SRAD_{mono}/100)$

Summarized, the modeling approach rested not upon introducing a competition sub-model. Instead, the climate and microclimate parameters were modified. Subsequently, the model and the genetic coefficients were evaluated and calibrated for the monocropping system and further used for the intercropping system. The coefficient of determination (R^2) and the root mean square error (RMSE) were used to estimate the variation between simulated and observed values.

Results of Wheat Modeling Experiments

Wheat benefited from being intercropped with maize. Grain yield of intercropped wheat was around 3 t ha^{-1} higher in comparison to monocropped wheat due to a higher number of tillers. The increased tiller number led to an increased dry matter yield of intercropped wheat of about 8 t ha^{-1} in comparison to monocropped wheat. The thousand kernel weight (TKW) in intercropped wheat was slightly reduced (Table 5).

Scenario 1: Modifying Light Interception

According to most other intercropping models, the DSSAT approach started with modifying the solar radiation as it was expected to be the competition factor with the highest influence on crop performance within intercropping systems. A similar approach to the turbid layer medium analogy was used but instead of using a submodel for solar radiation, the weather file in total was changed according to the determined shading pattern algorithms described before (3a) and evaluated throughout measured data.

Table 5: Mean comparisons of grain yield, TKW, tiller number and dry matter of intercropped and monocropped wheat in 2007/08

Treatment	Grain yield (kg [dm] ha^{-1})	TKW (g)	Tiller number (no./m²)	Dry matter (kg [dm] ha^{-1})
border rows within a plot	12400[a]	31[b]	864[a]	24506[a]
inner rows within a plot	9054[b]	34[a]	630[b]	16854[b]

[a,b] letters indicate significant differences between borderline (intercropping) and monocropping subplots at α = 0.05

The monocropping model used the standard weather file and an initial N input of 55 kg N ha^{-1} according to the calculated N surplus due to sugar beet as previous crop. The intercropping model used the modified solar radiation (3a) weather file without changing or modifying any other input parameters like soil, management and cultivar specifics. The results of model evaluation for the monocropped wheat showed a good fit between simulated and observed grain yield as well as dry matter accumulation even though the model did not simulate the slope of the sigmoid curve for grain yield adequately. The R^2 value was 0.96 for grain yield and 0.97 for dry matter yield respectively. Measured grain yield was 9.1 t ha^{-1}, simulated grain yield was 9.0 t ha^{-1}. Measured dry matter yield was 16.9 t ha^{-1}, simulated dry matter yield was 16.7 t ha^{-1}. The RMSE for grain yield at maturity was 49 kg ha^{-1}, the RMSE for dry matter accumulation at maturity was 127 kg ha^{-1}, indicating a model error of 0.54% and 0.75% respectively.

Running the intercropping model with the modified solar radiation alone only 10.3 t ha^{-1} grain yield was simulated instead of 12.4 t ha^{-1} according to the collected dataset (Table 5). Although the additional solar radiation obtained by the border rows accounted for 1.3 t ha^{-1} additional grain yield, a yield gap of approximately 2 t ha^{-1} remains still unexplained indicating that the increased grain yield could not only be explained by the higher solar radiation wheat border rows obtained until the maize reached a height of 1.4 m and shading occurred up to beginning of July. Similar observations were made for dry matter accumulation. Instead of the measured 24.5 t ha^{-1}, DSSAT simulated only 19.4 t ha^{-1}. The dry matter yield difference between monocropped and intercropped plots was 7.7 t ha^{-1}, but only 2.5 t ha^{-1} additional dry matter could be explained by the higher amount of incoming solar radiation during three-quarters of the growing season. The R^2 value was 0.99 for grain yield and 0.97 for dry matter yield. The RMSE for grain yield at maturity was 2071 kg ha^{-1}, the RMSE for dry matter accumulation at maturity was 5083 kg ha^{-1}.

Scenario 2: Modifying Light Interception and Soil Albedo

As differences in obtained sunlight could not fully explain the yield gap between monocropped and intercropped subplots, other parameters seemed to have an influence on crop performance. During the growing season there was no water stress so that differences in water availability could be excluded. Between April and August 2008, during co-existence of wheat and maize, there was 450 mm rainfall within 4.5 months. From the end of May until the beginning of July, the average soil moisture (30 cm soil depth) under maize was 48 vol % in both, monocropping and intercropping plots with the minimum value of 36 vol % under monocropped maize in beginning of July and the maximum value of 52 vol % under monocropped maize in beginning of June. In addition, the plots were sown on a relative homogenous field, so that great differences in spatial and temporal yield variability due to soil inhomogeneity dropped, too. However, because of the increased soil temperature and the increase in bare soil placements within the first rows it was assumed that there might have been differences in soil albedo. In the model the reflectance of the solar radiation from the soil surface accounts for potential evaporation from the soil surface. Whereas in the monocropped model the soil albedo value was set to 0.13, in the intercropped version it was reduced to 0.10.

Running the model showed that soil albedo had no great influence on crop performance. Grain yield at maturity increased only 7 kg ha^{-1} and dry matter yield at maturity increased 11 kg ha^{-1} respectively. Adjusted soil albedo in conjunction with sufficient soil water availability as an indicator for potential evaporation slightly changed performance of plants along field boundaries but could be neglected.

Scenario 3: Modifying Light Interception, Soil Albedo and Initial Conditions
At the field boundary, from sowing date until the sowing of maize in May, wheat faced no interspecific competition and less intra-specific competition wherefore microclimate changed. Most obvious, incoming solar radiation and top soil temperature increased. In addition wheat had a higher agressivity potential in comparison to maize. Hence, the mineralization of nitrogen was favoured and intercropped wheat got more nitrogen than its monocropped equivalent. Modifying the intercropped model once more, an additional initial N input of 190 kg N ha^{-1} (N_{min} + sugar beet leaves + 40 kg N ha^{-1} surplus) was given, taking the increased mineralization and the competitiveness of wheat into account. The initial condition for intercropped wheat was without limited nitrogen.

As a result, the intercropping model fitted well with R^2 values of 0.98 for dry matter yield and 0.96 for grain yield (Figure 3). The measured dry matter yield was 24.5 t ha^{-1}, the simulated 23.4 t ha^{-1}. The measured as well as the simulated grain yield was 12.4 t ha^{-1}. Wheat border rows used the increased sunlight, but foremost solar radiation in addition to increased N availability could adequately explain the grain yield increase. In comparison to other model approaches, where only solar radiation was taken into account as competition factor, DSSAT showed that other factors have to be regarded when intercropping should be simulated adequately. Otherwise, one competition factor is overestimated. A similar setting was given not only for the grain yield but also for dry matter yield. Running the model without additional N supply and solely with the modified solar radiation input, explained 3.4 t ha^{-1} additional dry matter yield. By contrast, running the model with additional N supply, dry matter yield increased to about 6.5 t ha^{-1}. The RMSE for grain yield at maturity was 2 kg ha^{-1}, the RMSE for dry matter accumulation at maturity was 1146 kg ha^{-1}, indicating a model error of 0.02% and 4.68% respectively.

Results of Maize Modeling Experiments

Maize intercropped with wheat suffered at the beginning of the growing season because of the competitiveness of wheat and reacted with less plant height (~ 0.5 m) and less dry matter accumulation (Table 7). But as wheat was harvested around three months earlier than maize, maize showed a recovery-compensation growth, already described by Li et al (2001b), resulting in maize borderline yields (rows at plot borders) at least as high as maize in monocropping (rows in plot centre).

The monocropped model was run taking the measured N_{min} values as well as the N surplus of sugar beet into account. In contrast, the intercropped model was run with N stress at the beginning of the growing season, so starting with less N content in the soil than the monocropped equivalent in order to take into account the increased competitiveness of wheat for N acquisition in comparison to maize. For example, the nitrogen withdrawal from wheat straw from the border rows (76 kg N ha^{-1}) was two times higher than the nitrogen withdrawal from monocropped wheat subplots (35 kg N ha^{-1}). Due to the recovery-compensation growth, the simulation of both monocropped and intercropped maize should be similar to each other (Figure 4), although the intercropping model was run with the modified solar radiation file. The steep slope of grain filling could not be simulated adequately, nevertheless, the simulation of grain yield at maturity showed a good fit between measured and observed grain yield. The R^2 value for monocropped maize was 0.83; the R^2 value for intercropped maize was 0.90. Measured grain yield of monocropped maize was 8.9 t ha^{-1}, simulated grain yield was 8.9 t ha^{-1}. The RMSE for grain yield at maturity was 18 kg ha^{-1} for monocropped maize and the RMSE for mean yield was 2017 kg ha^{-1}. Measured grain yield of intercropped maize was 8.9 t ha^{-1}, simulated grain yield was overestimated with 9.1 t ha^{-1}. The RMSE for grain yield at maturity was 259 kg ha^{-1} for intercropped maize and the RMSE for mean yield was 1353 kg ha^{-1}.

Table 6: R^2, RMSE, mean observed and simulated values for grain and dry matter yield under different scenarios

Variable	Mean (obs.)	Mean (sim.)	R^2	RMSE
Scenario 1: Grain yield (kg ha^{-1}) Dry matter yield (kg ha^{-1})	9053 14309	6414 12331	0.99 0.97	3351 3192
Scenario 2: Grain yield (kg ha^{-1}) Dry matter yield (kg ha^{-1})	9053 14309	6419 12338	0.99 0.97	3347 3184
Scenario 3: Grain yield (kg ha^{-1}) Dry matter yield (kg ha^{-1})	9053 14309	7704 14667	0.96 0.98	2258 1571

Figure 3: Simulated and observed grain yield and dry matter yield of monocropped (♦) and intercropped (∇) wheat in 2007/08 using scenario 3 for intercropped wheat

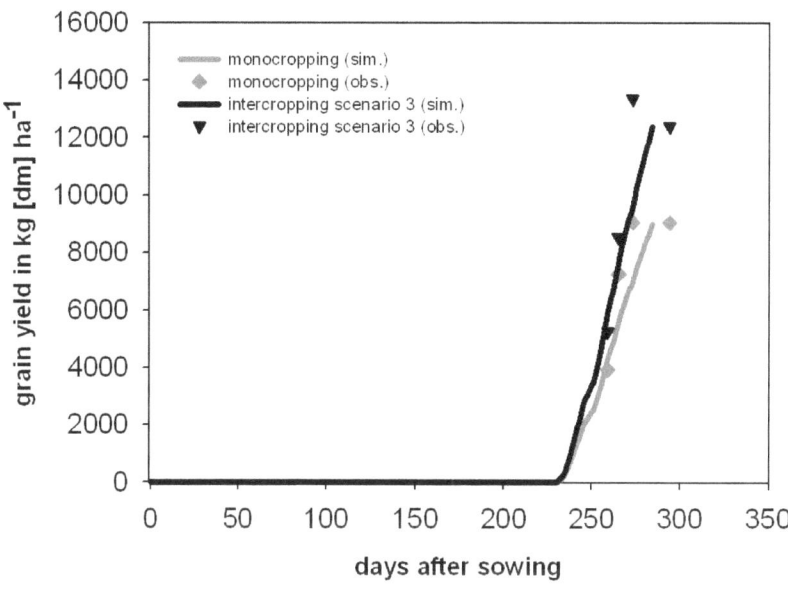

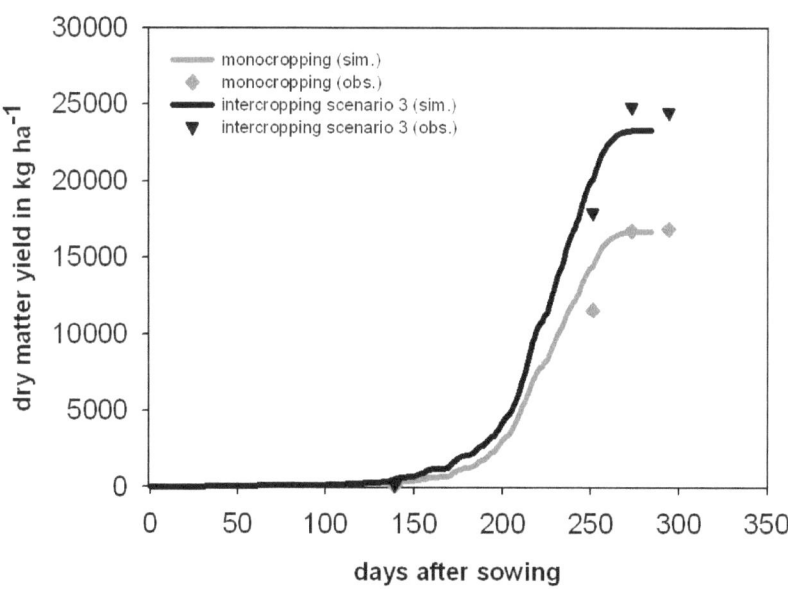

Figure 4: Simulated and observed grain yield of monocropped (♦) and intercropped (∇) maize in 2007/08

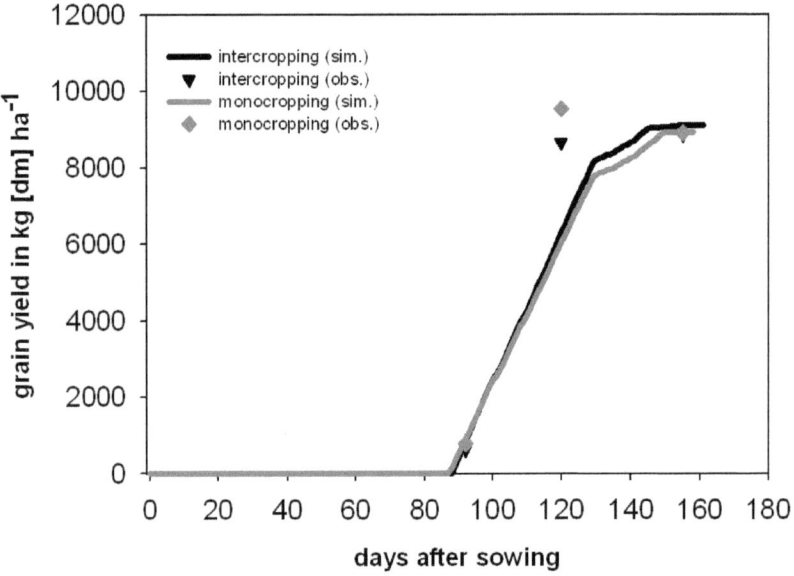

Table 7: Mean comparisons of grain yield, TKW, number of ears/plant, dry matter and LAI of intercropped and monocropped maize in 2007/08

Treatment	Grain yield (kg [dm] ha⁻¹)	TKW (g)	Ear number (no./plant)	Dry matter (kg [dm] ha⁻¹)	LAI
border rows within a plot	8852^a	261^a	1.4^a	14444^b	2.1^b
inner rows within a plot	8902^a	270^a	1.0^b	15554^a	2.9^a

^{a,b} letters indicate significant differences between borderline (intercropping) and monocropping subplots at α = 0.05

The evaluated shading pattern showed a strong shading of intercropped maize until beginning of July and a tanning up to July compared to its monocropped equivalent. Nevertheless, shading at the beginning and tanning at the end of the growing season were restricted to 80 and 20%, respectively, according to measurements. The steep slope of the linear shading pattern might have overestimated those dates. Maize as a C_4 plant could make a much better use of the increased solar radiation than wheat as a C_3 plant. Not the N supply, which was actually even lower for intercropped than for monocropped maize, was responsible for the recovery-compensation growth, but the effective usage of additional solar radiation after wheat harvest.

CONCLUSION

Most intercropping modeling studies were done in the 1990's and at the beginning of 2000. They mark a point where to start from. The various approaches seem to be promising as the validation of the divers models showed. Although a lot of crop growth and weed models have been used to simulate intercropping and interspecific competition, only a few show completely different approaches or ideas how to simulate competition in general. Most often, the turbid layer medium analogy in combination with a modified Beer's law was successfully used, but not applied in succession. Furthermore, the ability to model intercropping is still not introduced in most common process-oriented crop growth models; instead it is often introduced as stand-alone simulation approach. Modeling intercropping should take a step forward from pure modeling to scenario simulation in order to advice crop growers or to improve cropping systems. Park et al. (2003) concluded in a similar way concerning crop-weed models: it is necessary to increase the knowledge of spatial and temporal variability in model parameters if the models' usage should be extended to be more predictive and advisory.

The more statically and empirical models – often no-named – are restricted to model a special effect or phenomenon of intercropping for research use in the main. In contrast, dynamic and mechanistic models which take soil-plant-atmosphere interactions into account like APSIM, INTERCOM and STICS build up an applicative base useful not only for researchers but also for advisers and extension services. Nevertheless, in European, North-American and Australian agriculture, intercropping or smallholder farming is less practiced whereas most models have been evolved in those countries. In addition, modeling the various intercropping systems – and not only maize and legume systems - requires models which include a great number of crops. With 16 integrated species, DSSAT offers a possibility to fulfil those basic requirements. As it is a generic model, other crop models or species can be introduced into DSSAT much easier.

Intercropping is still an up-to-date topic. It is practiced not only in smallholder farming in countries like Africa, India or China, but also in the US. The Alternative Agriculture News titled "Contour strip intercropping can reduce erosion and energy costs" (AANews, March 1994). The American farmer Paul Mugge, who practices strip intercropping with maize and soybean, stated that being profitable, being efficient in the terms of resources, understanding more about ecology and using that understanding (Kendall, 1996, 1997) are his driving forces for doing intercropping. In a comprehensive analysis of our agricultural systems, we should not only examine yields, but also the cost of the inputs used to obtain them. Intercropping may help eliminate unnecessary use of non-renewable resources in modern agriculture or it may help using these resources more efficiently (Horwith, 1985, p. 289). Models should be extended to more diversified intercropping systems, to a greater variability within countries and to other aspects of intercropping than yield or plant performance, e.g. quality aspects (Andrighetto et al., 1992; Dawo et al., 2007). In addition, the modeling of case studies should be broadened to the modeling and simulating of intercropping scenarios to either study the sustainability potential (e.g., reduced nitrate leaching, reduced fertilizer input) more closely or to adjust and further improve existing cropping systems. That could be done in an inverse way to our model scenario approach: the DSSAT model showed that within an intercropping system of maize and wheat, wheat benefits from its increased nitrogen availability in this system. Hence, less fertilizer may be applied to wheat, reducing leaching and high input costs without reducing yield. Nitrogen scenario simulation could be a possibility to improve the cropping system.

Modeling intercropping with regard to field boundary cultivation broadens the view for unconscious intercropping because of small field size and for patchy agricultural landscapes. There, the amount of field boundaries becomes important and relevant for yield expectance and cultivation practice and hence, for sustainability. In order to model intercropping with special regard to field boundary cultivation, the new model approach with the DSSAT crop growth model showed the possibility to simulate general competition and beneficial effects without introducing a submodel. Instead, shading algorithms were evaluated to modify microclimate and microclimate changes such as incoming solar radiation within field boundaries. In addition, increased soil temperature implicating an increased mineralization and hence a higher N availability for wheat border rows could be taken into account. It is important to consider more than one competition factor like only solar radiation, because species use sunlight more or less efficiently. Different solar radiation amounts could not explain yield differences in all cases. For intercropped maize, solar radiation was an important competition factor. Hence, the recovery-compensation growth could be modeled adequately, because maize could use the increased solar radiation after wheat harvest efficiently even if there was strong shading at the beginning of the maize growing period. In contrast, wheat benefited from the increased solar radiation until flowering but benefited even more from the increased N availability.

ACKNOWLEDGMENT

The authors' research topic is embedded in the International Research Training Group of the University of Hohenheim and China Agricultural University, entitled "Modeling Material Flows and Production Systems for Sustainable Resource Use in the North China Plain". We thank the German Research Foundation (DFG) and the Ministry of Education (MOE) of the People's Republic of China for financial support.

REFERENCES

Adiku, S. G. K., Carberry, P. S., Rose, C. W., McCown, R. L., & Braddock, R. (1995). A maize (*Zea mays*)-cowpea (*Vigna unguiculata*) intercrop model . In *H. Sinoquet & P. Cruz (Es.), Ecophysiology of Tropical Intercropping* (pp. 397–406). Paris: INRA editions.

Aikman, D. P., & Benjamin, L. R. (1994). A model for plant and crop growth, allowing for competition for light by the use of potential and restricted projected crown zone areas. *Annals of Botany, 73*, 185–194. doi:10.1006/anbo.1994.1022

Alternative Agriculture News. (1994). Contour strip intercropping can reduce erosion. *AANews*.

Andrighetto, I., Mosca, G., Cozzi, G., & Berzaghi, P. (1992). Maize-soybean intercropping: effect of different variety and sowing density of the legume on forage yield and silage quality. *Journal of Agronomy andCcrop Science, 168*(5), 354-360.

Ball, D. A., & Shaffer, M. J. (1993). Simulating resource competition in multispecies agricultural plant communities. *Weed Research, 33*, 299–310. doi:10.1111/j.1365-3180.1993.tb01945.x

Baumann, D. T., Bastiaans, L., Goudriaan, J., Van Laar, H. H., & Kropff, M. J. (2002). Analysing crop yield and plant quality in an intercropping system using an eco-physiological model for interplant competition. *Agricultural Systems, 73*, 173–203. doi:10.1016/S0308-521X(01)00084-1

Berntsen, J., Hauggaard-Nielsen, H., Olesen, H., Petersen, B. M., Jensen, E. S., & Thomsen, A. (2004). Modelling dry matter production and resource use in intercrops of pea and barley. *Field Crops Research, 88*(1), 59–73. doi:10.1016/j.fcr.2003.11.012

Brisson, N., Bussiére, F., Ozier-Lafontaine, H., Tournebize, R., & Sinoquet, H. (2004). Adaptation of the crop model STICS to intercropping. Theoretical basis and parameterisation. *Agronomie, 24*, 409–421. doi:10.1051/agro:2004031

Caldwell, R. M. (1995). Simulation models for intercropping systems . In *H. Sinoquet & P. Cruz (Es.), Ecophysiology of Tropical Intercropping* (pp. 353–368). Paris: INRA editions.

Carberry, P. S., McCown, R. L., Muchow, R. C., Dimes, J. P., Probert, M. E., Poulton, P. L., & Dalgliesh, N. P. (1996). Simulation of a legume ley farming system in northern Australia using the agricultural production systems simulator. *Australian Journal of Experimental Agriculture, 36*, 1037–1048. doi:10.1071/EA9961037

Dawo, M. I., Wilkinson, J. M., Sanders, F. E. T., & Pilbeam, D. J. (2007). The yield of fresh and ensiled plant material from intercropped maize (*Zea mays*) and beans (*Phaseolus vulgaris*). *Journal of the Science of Food and Agriculture, 87*, 1391–1399. doi:10.1002/jsfa.2879

Federer, W. T. (1993). *Statistical design and analysis for intercropping experiments: Vol. 1. Two crops*. New York: Springer.

García-Barrios, L., Mayer-Foulkes, D., Franco, M., Urquijo-Vásquez, G., & Franco-Pérez, J. (2001). Development and validation of a spatially explicit individual-based mixed crop growth model. *Bulletin of Mathematical Biology, 63*, 507–526. doi:10.1006/ bulm.2000.0226

Grant, R. F. (1992). Simulation of competition among plant populations under different managements and climates.

Grant, R. F. (1994). Simulation of competition between barley (*Hordeum vulgare L.*) and wild oat (*Avena fatua L.*) under different managements and climates. *Ecological Modelling, 71*, 269–287. doi:10.1016/0304-3800(94)90138-4

Horwith, B. (1985). A role for intercropping in modern agriculture. *Bioscience, 35*(5), 286–290. doi:10.2307/1309927

Inal, A., Gunes, A., Zhang, F., & Cakmak, I. (2007). Peanut/maize intercropping induced changes in rhizosphere and nutrient concentrations in shoots. *Plant Physiology and Biochemistry, 45*, 350–356. doi:10.1016/j.plaphy.2007.03.016

Jensen, E. S. (2006). *Intercrop; Intercropping of cereals and grain legumes for increased production, weed control, improved product quality and prevention of N-losses in European organic farming systems* (Rep. No. QLK5-CT-2002-02352. Risø. Jolliffe, P. A. (1997). Are mixed populations of plant species more productive than pure stands? *Acta Oecologica Scandinavica, 80*(3), 595–602.

Jones, J. W., Hoogenboom, G., Porter, C. H., Boote, K. J., Batchelor, W. D., & Hunt, L. A. (2003). The DSSAT cropping system model. *European Journal of Agronomy, 18*, 235–265. doi:10.1016/S1161- 0301(02)00107-7

Kendall, J. (1996/1997). PFI profile: Paul and Karen Mugge. *The practical farmer, 11*(4).

Kiniry, J. R., & Williams, J. R. (1995). Simulating intercropping with the ALMANAC model . In Sinoquet, H., & Cruz, P. (Eds.), *Ecophysiology of Tropical Intercropping* (pp. 387–396). Paris: INRA editions.

Kiniry, J. R., Williams, J. R., Gassman, P. W., & Debaeke, P. (1992). A general process-oriented model for two competing plant species. *Transactions of the ASAE. American Society of Agricultural Engineers, 35*(3), 801–810.

Koeller, K. H., & Linke, C. (2001). *Erfolgreicher Ackerbau ohne Pflug: Wissenschaftliche Ergebnisse – Praktische Erfahrungen.* Frankfurt, Germany: DLG.

Kropff, M. J., & Spitters, C. J. T. (1992). An eco-physiological model for interspecific competition, applied to the influence of *Chenopodium album L.* on sugar beet. I. Model description and parameterization. *Weed Research, 32*, 437–450. doi:10.1111/j.1365-3180.1992.tb01905.x

Kropff, M. J., & van Laar, H. H. (1993). *Modelling crop-weed interactions.* CAB International, in association with the International Rice Research Institute.

LEL Schwäbisch Gmünd, LTZ Augustenberg, LVVG Aulendorf, LSZ Boxberg, LVG Heidelberg & HuL Marbach. (2007). *Stammdatenblätter Landwirtschaft „Nährstoffvergleich Feld-Stall.* Tabelle 5a.

Li, L., Sun, J., Zhang, F., Li, X., Rengel, Z., & Yang, S. (2001). Wheat/maize or wheat/soybean strip intercropping II. Recovery or compensation of maize and soybean after wheat harvesting. *Field Crops Research, 71*, 173–181. doi:10.1016/S0378- 4290(01)00157-5

Li, L., Sun, J., Zhang, F., Li, X., Yang, S., & Rengel, Z. (2001). Wheat/maize or wheat/soybean strip intercropping I. Yield advantage and interspecific interactions on nutrients. *Field Crops Research, 71*, 123–137. doi:10.1016/S0378-4290(01)00156-3

Lowenberg-De Boer, J., Krause, M., Deuson, R., & Reddy, K. C. (1991). Simulation of yield distributions in millet-cowpea intercropping. *Agricultural Systems, 36*, 471–487. doi:10.1016/0308-521X(91)90072-I

Nelson, R. A., Dimes, J. P., Paningbatan, E. P., & Silburn, D. M. (1998). Erosion/productivity modelling of maize farming in the Philippine uplands part I: Parameterising the agricultural production systems simulator. *Agricultural Systems, 58*(2), 129–146. doi:10.1016/S0308-521X(98)00043-2

O'Callaghan, J. R., Maende, C., & Wyseure, G. L. C. (1994). Modelling the intercropping of maize and beans in Kenya. *Computers and Electronics in Agriculture, 11,* 351–365. doi:10.1016/0168-1699(94)90026-4

Ozier-Lafontaine, H., Bruckler, L., Lafolie, F., & Cabidoche, Y. M. (1995). Modelling root competition for water in mixed crops: a basic approach . In Sinoquet, H., & Cruz, P. (Eds.), *Ecophysiology of Tropical Intercropping* (pp. 189–187). Paris: INRA editions.

Ozier-Lafontaine, H., Lafolie, F., Bruckler, L., Tournebeze, R., & Mollier, A. (1998). Modelling competition for water in intercrops: theory and comparison with field experiments. *Plant and Soil, 204,* 183–201. doi:10.1023/A:1004399508452

Park, S. E., Benjamin, L., & Watkinson, A. R. (2003). The theory and application of plant competition models: an agronomic perspective. *Annals of Botany, 92,* 741–748. doi:10.1093/aob/mcg204

Reisch, E., & Knecht, G. (1995). *Betriebslehre.* Stuttgart, Germany: Ulmer.

Rossiter, D. G., & Riha, S. J. (1999). Modeling plant competition with the GAPS object-oriented dynamic simulation model. *Agronomy Journal, 91*(5), 773–783. doi:10.2134/agronj1999.915773x

SAS Institute. (2009). *The SAS System for Windows (Release 9.2).* Cary, NC: SAS Institute.

Sellami, M. H., & Sifaoui, M. S. (1999). Modelling solar radiative transfer inside the oasis; Experimental validation. *Journal of Quantitative Spectroscopy & Radiative Transfer, 63,* 85–96. doi:10.1016/S0022- 4073(98)00137-X

Sinoquet, H., & Caldwell, R. M. (1995). Estimation of light capture and partitioning in intercropping systems . In Sinoquet, H., & Cruz, P. (Eds.), *Ecophysiology of Tropical Intercropping* (pp. 79–97). Paris: INRA editions.

Sinoquet, H., Rakocevic, M., & Varlet-Grancher, C. (2000). Comparison of models for daily light partitioning in multispecies canopies. *Agricultural and Forest Meteorology, 101,* 251–263. doi:10.1016/ S0168-1923(99)00172-0

Song, Y. N., Marschner, P., Li, L., Bao, X. G., Sun, J. H., & Zhang, F. S. (2007). Community composition of ammonia-oxidizing bacteria in the rhizosphere of intercropped wheat (*Triticum aestivum* L.), maize (*Zea mays* L.) and faba bean (*Vicia faba* L.). *Biology and Fertility of Soils, 44*(2), 307–314. doi:10.1007/ s00374-007-0205-y

Thornton, P. K., Dent, J. B., & Caldwell, R. M. (1990). Applications and issues in the modelling of intercropping systems in the tropics. *Agriculture Ecosystems & Environment, 31*(2), 133–146. doi:10.1016/0167-8809(90)90215-Y

Tinsley, R. L. (2004). *Developing smallholder agriculture – a global perspective.* Brussels, Belgium: AgBé Publishing.

Tsubo, M., & Walker, S. (2002). A model of radiation interception and use by maize-bean intercrop canopy. *Agricultural and Forest Meteorology, 110,* 203–215. doi:10.1016/S0168-1923(01)00287-8

Tsubo, M., Walker, S., & Ogindo, H. O. (2005). A simulation of cereal-legume intercropping systems for semi-arid regions. I. Model development. *Field Crops Research, 93,* 10–22. doi:10.1016/j. fcr.2004.09.002

Vandermeer, J. (1989). *The ecology of intercropping.* New York: Cambridge University.

Wiles, L. J., & Wilkerson, G. G. (1991). Modeling competition for light between soybean and broadleaf weeds. *Agricultural Systems, 35,* 37–51. doi:10.1016/0308-521X(91)90145-Z

Yokozawa, M., & Hara, T. (1992). A canopy photosynthesis model for the dynamics of size structure and self-thinning in plant populations. *Annals of Botany, 70*, 305–316.

Zhang, F., & Li, L. (2003). Using competitive and facilitative interactions in intercropping systems enhances crop productivity and nutrient-use efficiency. *Plant and Soil, 248*, 305–312. doi:10.1023/A:1022352229863

Zhang, L. (2007). *Productivity and resource use in cotton and wheat relay intercropping. Chapter 6: Development and validation of SUCROS-Cotton: A mechanistic crop growth simulation model for cotton, applied to Chinese cropping conditions.* Unpublished doctoral dissertation, Wageningen University, The Netherlands.

6 Chapter III:
Extension and evaluation of intercropping field trials using spatial models

PUBLICATION III:

Knörzer, H., Müller, B.U., Guo, B., Graeff-Hönninger, S., Piepho, H.-P., Wang, P., and Claupein, W. (2010): Extension and evaluation of intercropping field trials using spatial models. Agronomy Journal 102 (3), pp. 1023-1031.

After reviewing and studying a wide range of publications dealing with intercropping, a basic issue in designing and later on analyzing intercropping experiments was identified. At least row, strip, and relay intercropping experiments contravene one of the three fundamental statistical principles as there are blocking, replication, and randomization. The latter cannot be fulfilled as row and strip intercropping systems require alternating rows or strips. Thus, spatial variability has not been taken into consideration so far while analyzing row or strip intercropping systems with a common analysis of variance. As a result, detected significant differences between treatments or traits such as yield or yield components bear the potential of being estimated in a too optimistically manner. The application of spatial models for analyzing row and strip intercropping trials and for improving the model fit are suggested. In chapter III, several spatial models were presented, tested, and applied for the German and Chinese field experiments during 2007 and 2009. In addition, the question arose, if small fields within fragmented agricultural landscapes, like parts of China, are intercropping in a larger scale? Intercropping is based upon field border effects, and hence, the sum of field borders becomes important within those landscapes when analyzing crop performance. Again, the spatial dimension of intercropping has to be considered and is shown in the following chapter.

The publication of this chapter had been announced for the Research Highlight Program 2010 of the American Society of Agronomy, the Crop Science Society of America, and the Soil Science Society of America (ASA-CSSA-SSSA).

Extension and Evaluation of Intercropping Field Trials Using Spatial Models

H. Knörzer, S. Graeff-Hönninger, and W. Claupein, Univ. of Hohenheim, Institute of Crop Science, Fruwirthstrasse 23, 70593 Stuttgart, Germany;
B.U. Müller and H.-P. Piepho, Univ. of Hohenheim, Bioinformatics, Fruwirthstrasse 23, 70593 Stuttgart, Germany;
B. Guo and P. Wang, China Agricultural Univ., College of Agronomy and Biotechnology, No. 2 West Yuan Ming Yuan Rd., Beijing 100094, China.
H. Knörzer and B. U. Müller equally contributed to this work.

Article from the Agronomy Journal 102 (3):1023–1031 (2010), with permission, copyright © 2010 by the American Society of Agronomy.

ABSTRACT

Intercropping has oft en been considered as a secluded cropping system within one field. However, in African and Asian countries, where intercropping is widespread, the system can be looked on at a much larger scale: small fields alternate as strips with different crops grown on them, turning the collection of fields into a kind of unplanned intercropping. The more fragmented the agricultural landscape, the more relevant the borders can become. Traditionally, statistical analysis of intercropping experiments has been done by a simple analysis of variance without taking spatial models into account. But especially strip intercropping experimental arrangements lack in randomization as the cropping system imposes alternating strips. T us, spatial variability and its effect on yield have to be regarded differently. Two different features of intercropping experiments were studied in the present paper: statistical peculiarities of intercropping designs and the border effect which is a key component of intercropping performance. Field trial results from Germany and China indicated that intercropping showed significant border row effects within the first four rows. For statistical analysis, different spatial models were added to the baseline model to account for the spatial trend and to check whether or not standard models are suitable for analyzing intercropping experiments. The results showed that for the German experiment the baseline model fitted well in the year 2008 and a common analysis of variance seemed to be well suited. However, for the Chinese experiments and the German experiment in the year 2009 the spatial models improved the model fit.

INTRODUCTION

Ecology meets agronomy whenever there is a patchy agricultural landscape and whenever there is more than one crop grown simultaneously on the same (small) field. Smallholder farming and scarcity of land especially for African and Asian farmers on the one hand and consideration of sustainability, resource use efficiency, and yield stability for American and European farmers on the other hand, have led to a great diversity of intercropping systems, each of it with its individual and complex interactions of beneficial and competitive effects occurring between different neighboring species.

By definition intercropping is the growing of two or more crops within the same or an overlapping growing season within the same field. To gauge, to handle, or to analyze them is much more difficult in comparison to monocropping systems. Experiments dealing with row or strip intercropping lack in a basic scientific principle: they cannot be randomized. So how should interspecific competition effects be weighted among species, within fields or even patchy agricultural landscapes? Are common mixed models used for statistical analysis in monocropping scenarios applicable for intercropping? Or do we have to broaden our view on intercropping to a spatial dimension within landscapes as well as within statistics?

Intercropping has not been analyzed from the point of view that fragmented agricultural landscapes with predominant smallholder farming correspond to intercropping at a larger scale, defining the sum of small fields next to each other as intercropping. In Africa, for example, the average farm size is about 2 ha and 83% of all cropped land in northern Nigeria and 94% in Malawi are intercropping area; in China the average farm size is about 0.1 ha and 20 to 25% of arable land is intercropped area. In India, the average farm size is about 1.2 to 2.7 ha and 17% of arable land is intercropped (Beets, 1982; Cohen, 1988; FAO, 2006; Li, 2001; Li et al., 2007; Vandermeer, 1989; Wubs et al., 2005). From this point of view, intercropping could be subjected to a patch analysis where the sum of small fields in juxtaposition with each other are the individual patches dependent on the distance from one border to another. Hence, the theory of borders or boundaries from ecology can be transferred to agronomy. According to Fernandez et al. (2002), the ecological definition of a boundary reads as follows: The border and the two edges appear as a consequence of the interaction and constitute what is called a boundary. The terms field boundary, edge, and border effect define similar or even the same facts as well as circumstances and can be regarded as synonyms. For intercropping studies, the term border effect is preferred.

Statistical analysis of intercropping experiments has mostly been done by a simple ANOVA with intercropping arrangements handled as treatments (Agyare et al., 2006; Ghaffarzadeh et al., 1994). For example, when searching for the optimal row number or strip width, the number of rows was considered as treatment, neglecting the fact that especially strip intercropping experimental arrangements lack in randomization as the cropping system requires alternating strips. Different methods such as use of blocking and randomization were developed to reduce the effect of spatial trend, which is caused by weather, soil heterogeneity, and several other factors (Fisher, 1925; Edmondson, 2005). A key issue in the analysis of intercropping experiments is that randomization is impossible with respect to position and crop effects. Besides blocking and randomization, many spatial methods were proposed for adjusting for spatial trend (Bartlett, 1978; Wilkinson et al., 1983; Schwarzbach, 1984; Williams, 1986; Gilmour et al., 1997; Gleeson, 1997). A common feature of these methods is that plots that are closer together are assumed to have a higher correlation than plots farther apart. The precision of the estimates can be improved through advanced experimental designs and by use of spatial models.

A similar experiment with restricted randomization is the line-source sprinkler experiment described in SAS Institute (2009, p. 4049) and Piepho et al. (2004), where it is shown how spatial analysis can be used to account for the lack of randomization. Here, we follow a similar approach. Before describing the approach we should like to stress that spatial analysis is not a general substitute for randomization; the principle of randomization should be adhered wherever possible, because this is the best way to avoid biases (Cochran and Cox, 1957; John and Williams, 1995). But there are situations where for scientific reasons randomization is not possible with respect to a factor of interest, such as in intercropping, and in these situations spatial analysis may help. This should not be taken to mean that we can generally abandon the idea of randomization, relying on spatial modeling to fix the problem after the experiment (Bailey et al., 1995).

A spatial analysis is proposed in the following study with regard to the analysis of border effects in intercropping systems lacking in randomization. Our focus was on a one-dimensional analysis within plots, because along the strip direction a greater continuity (longer range or dependence length) is generally observed.

MATERIALS AND METHODS
Field Trial in China

Experiment 1 was conducted in northeast China (37.17/59.9″ N, 116.17/59.9″ E) at the China Agricultural University experimental station in Wuqiao, Hebei province, during the years 2008 and 2009. In Wuqiao, the average rainfall per year is 562 mm; the average temperature per year is 13.1.C. In this region, mainly alluvial soils occur with a disposition to salinization.

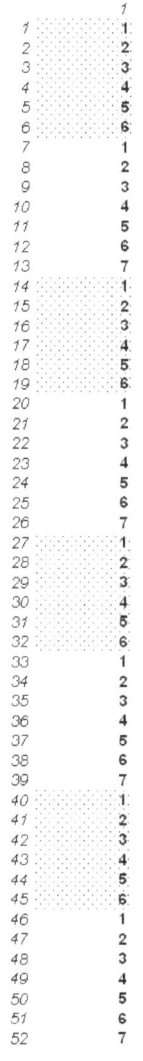

The experiment was designed in four nonrandomized complete blocks. The experimental layout is illustrated in Fig. 1. One intercropping system with maize (*Zea mays* L.) and peanut (*Arachis hypogaea* L.) was used for the Chinese experiment. Because the cropping system requires alternating strips, a randomization was not possible at all, even for the factor system. There was just a single system and a single column of plots with crops allocated to plots in an alternate pattern as usual in intercropping. By contrast, in the German experiment, there were two intercropping systems within each replicate, so that the factor system could be randomized.

Maize and peanut were sown in May with a row spacing of 60 and 30 cm, respectively. Both crops were sown in alternating plots 8 m wide and 20 m long. The Chinese experiment had one column, six strips for each plot of peanut, seven strips for each plot of maize, and in total 52 strips. Thus, the Chinese experiment had six position levels for peanut or seven position levels for maize, with strip one reflecting the plot border and strip six or seven the strip which is furthermost from the plot border.

Data of each strip within the plots were collected via square meter-cuts to detect differences within the positions of the plots. The term 'strip' under those circumstances is not automatically synonymous with crop row as it may indicate a unit covering several rows. Nevertheless, in the Chinese experiment, each strip comprised a single crop row. The plots were big enough to reflect monocropping within the central strips and each plot could be considered to be a small, secluded field.

At maturity, dry matter and grain yield of both crops were determined in 2008. In addition, in 2008 the number of kernels per ear and the thousand kernel weight (TKW) were measured for maize. In 2009, only data of grain yield for both crops were available.

Figure 1: Field layout of the field trial in China. The randomization of the systems and the levels of the position effect are illustrated in the field layout. Each rectangle is considered as plot. Levels of the position effect within the plots are bold faced

Field Trial in Germany

Experiment 2 was performed in southwest Germany (48.27/36˝ N, 8.33/36˝ E) at the University of Hohenheim experimental station 'Ihinger Hof' during 2007/2008 and 2008/2009. The average rainfall per year is around 690 mm with an average temperature of 7.9.C. The soils are mainly keuper with loess layers.

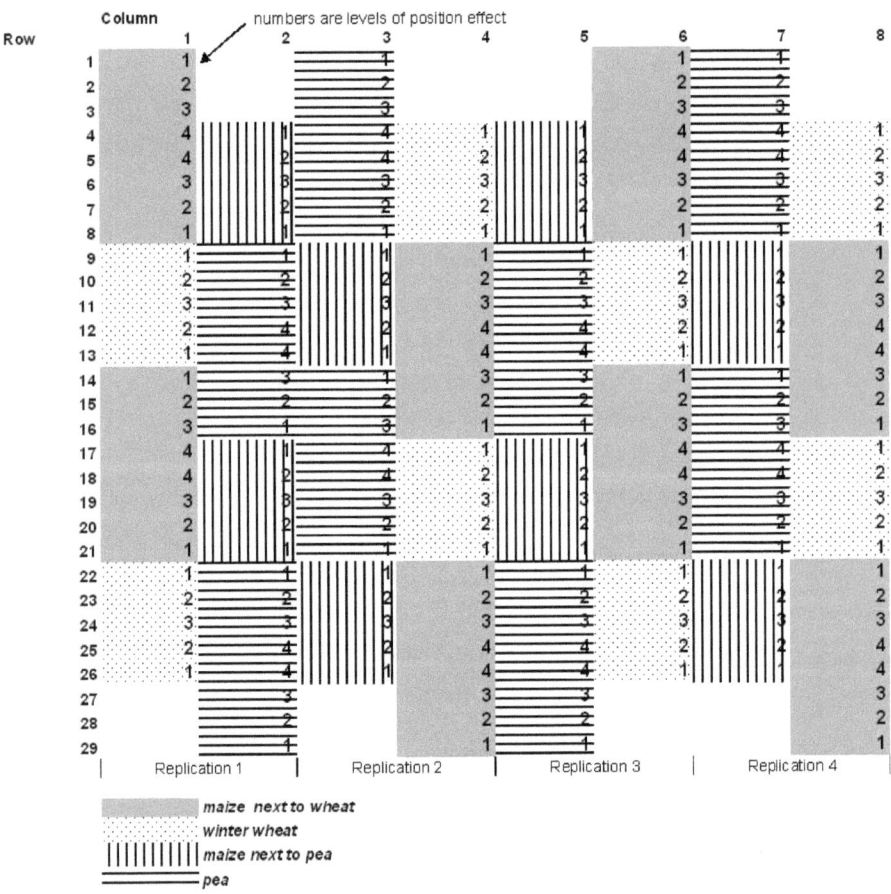

Figure 2: Field layout of the field trial in Germany. The randomization of the systems and the levels of the position effect are illustrated in the field layout. Each rectangle is considered as plot

The experiment comprised two intercropping systems: maize/pea (*Pisum sativum* L.) and maize/wheat. The field layout is depicted in Fig. 2. The two systems were randomized in four complete blocks of two columns each. Each replicate consisted of two columns with eight plots each. The two species were planted in an alternate pattern. Thus, randomization was restricted to the factor system, whereas for crop species no randomization was possible. Each plot was 10 by 10 m^2 for wheat and pea, and 12 by 10 m^2 for maize, respectively, and included five strips (5 by 2 by 10 m^2) for wheat and pea, and eight strips (8 by 1.5 by 10 m^2) for maize. One experimental unit - denoted "strip" in the statistical analysis hereafter - of wheat and pea contained 15 crop rows and two maize rows. Thus, the German experiment consisted of five strips with three position levels for wheat and pea: the distance 0 to 2 m from the border occurred two times as well as the distance 2 to 4 m. The distance 4 to 6 m reflected a monocropping situation. In addition, the maize experiment had eight strips with four position levels: rows 1 and 2, rows 3 and 4, and rows 5 and 6 from plot border occurred two times within each plot. Rows 7 and 8 reflected monocropping.

The collection of plots in total could be considered as small fields in a patchy agricultural landscape (macro level). To study the effects of that patchiness on yield distribution within the field/plot, data was collected within the plot strips (micro level). Strips were oriented north-south. Within each strip data were collected via square meter-cuts to detect differences between the positions within the plots.

Wheat was sown in October 2007 and 2008 with a row spacing of 13 cm and a plant density of 300 plants per m^2. Pea was sown at the end of March in both 2008 and 2009. Row spacing was 13 cm and plant density 70 plants per m^2. Maize was sown in May 2008 and 2009 with a row spacing of 75 cm and a plant density of 10 plants per m^2. During the growing season, neither water nor N stress occurred so that differences in plant growth and yield performance could be attributed to intra- and interspecific competition.

At the beginning and the end of the growing season, the number of plants per m^2 was counted to determine the stand density. During the growing season, leaf area index (LAI) was measured destructively with an LI-3100 Area Meter (LI-COR, Lincoln, NE) on a weekly basis by harvesting several plants of each strip. After maturity, square meter-cuts were taken to determine dry matter, grain yield, and TKW.

Statistical Methods

To detect significance of response differences between strips with different distances from the field border, a mixed model with fixed position effects and a random effect *t* modeling the spatial trend by covariance structures was fitted for different traits, grain yield, TKW, LAI, stand density, and kernels/ear (Schabenberger and Gotway, 2005). Each species was analyzed separately in this study. The position effect models the response as a function of the distance of the strips to the border of the plot. The plots, therefore, had different levels for the position effect based on distance of the strips to the border. The position effect is illustrated in Fig. 1 and 2.

The baseline model for a single plot of strips pertaining to the same crop species can be described by:

$$y = \mu + \beta + \varepsilon \quad [1]$$

where y is the observed yield for each strip, μ is an intercept (column effect), β is the fixed effect for the position, which models the distance of each strip within the plot to the border of the plot, and ε is the error effect, assumed to be independently normally distributed with zero mean and variance σ^2. The position effect had different levels for the German and Chinese experiment to detect differences between the strips of the plots to the border of the plot. For the German experiment four levels of the position effect were assumed for maize, and three levels for wheat and pea. Six levels were assumed for peanut and seven levels for maize in China. For the German experiment also a fixed effect for the plot was fitted explicitly because there were several plots within the replicates.

Different spatial models were added on to the baseline model to account for the spatial trend within plots (Piepho et al., 2008; Müller et al., 2010). Spatial trend was considered to be evident across the whole field. Therefore, it was assumed that most of the spatial trend is found between strips within plots. The full spatial model for a single plot of strips can be written as:

$$y = \mu + \beta + t + \varepsilon \quad [2]$$

where t is the spatial trend effect. The error effect ε is denoted in spatial modeling as nugget (Piepho et al., 2008). The spatial models differ in the way the spatial trend effect t is modeled. All spatial models assume that trend effects of strips in the same plots are correlated in such a way that covariance decays with distance among strips.

A spatial model for local trend t used in this analysis was the linear variance (LV) model proposed by Williams (1986). This model assumes linearly decreasing correlations among plots with increasing distance. Additional models for local trend t used in this study were the first-order autoregressive AR(1) model (Cullis and Gleeson, 1991; Gilmour et al., 1997; Gleeson and Cullis, 1987), the exponential (Exp) model, the Gaussian (Gau) model, and the spherical (Sph) model (Schabenberger and Gotway, 2005). The spatial trend of the different models can be described by the following equations with $|j_1 - j_2|$ denoting the distance between the j_1^{th} and j_2^{th} plots of a column and t_j the trend on the jth plot:

LV: $\text{cov}(t_{j_1}, t_{j_2}) = \eta - \Phi |j_1 - j_2|$ [3]

AR(1): $\text{cov}(t_{j_1}, t_{j_2}) = \sigma^2 \rho^{|j_1 - j_2|}$ [4]

Exponential: $\text{cov}(t_{j_1}, t_{j_2}) = \sigma^2 [\exp(-|j_1 - j_2|/d)]$ [5]

Gaussian: $\text{cov}(t_{j_1}, t_{j_2}) = \sigma^2 [\exp(-|j_1 - j_2|/d^2)]$ [6]

Spherical: $\text{cov}(t_{j_1}, t_{j_2}) = \sigma^2 [1 - (1.5 (|j_1 - j_2|)/d - 0.5(|j_1 - j_2|/d^3)]$ if $0 < |j_1 - j_2| \leq d$ [7]

All these models were fitted with and without a nugget effect ε. All models were fitted by REML using the MIXED procedure of the SAS System 9.2 (SAS Institute, 2009). Models which have the same fixed effect were compared using Akaike Information Criterion (AIC) (Wolfinger, 1996), defined as minus twice the REML log-likelihood plus twice the number of variance parameters. The smaller the value of AIC the better the model fit.

The differences between the positions of the strips to the border of the plot were subjected to a t-test. A letter display of all pairwise comparison was generated (Piepho, 2004). For comparison, we also performed a Tukey-Kramer-test.

RESULTS

Field Border Effect sine qua non in Intercropping

Results from strip samplings showed that there is a patterned yield distribution within the plots. Crop performance is influenced by plot borders or the distance from plot border and the kind of neighboring species (Fig. 3).

Field trials with wheat and pea in Germany: Wheat intercropped with maize had a significantly increased grain yield within the first 2 m (12 t [dm] ha^{-1}) in comparison to the 2 to 4 m and the 4 to 6 m (9 t [dm] ha^{-1}) distance from border strip in 2008 and in 2009. Grain yield differences were approximately 3 t [dm] ha^{-1} in 2008/2009 and dry matter yield differences were 7 t ha^{-1} in 2008/2009 with intercropped wheat removing 120 kg ha^{-1} more N from the soil than monocropped wheat until dough ripe stage in 2008 and 2009. The increased grain yield was mainly due to the increased number of tillers of intercropped wheat which had 864 tillers per m^2 in 2008. On average, intercropped wheat had 234 tillers per m^2 more than monocropped wheat whereas the TKW was similar with 34 g in intercropped and 35 g in monocropped wheat. Yield and yield components differences were based on the first four rows. The complete 2 to 10 m strips were harvested on the one hand with a harvester and on the other hand with square meter-cuts of only a few rows. Both harvesting strategies showed the same significant differences (results not shown), but the differences were more distinctive when analyzing row-by-row harvests, indicating that the yield increase was mainly due to higher yield within the first four rows.

Pea grain yield showed no significant differences between the different strips or distances between border strips and inner strips. Grain yield of intercropped pea in 2008 was 5 t [dm] ha^{-1}, grain yield of monocropped pea was 4 t [dm] ha^{-1}. In 2009, intercropped and monocropped pea had the same grain yield of 5 t [dm] ha^{-1}. Intercropped as well as monocropped pea had 10 pods per plant and 3.4 kernels per pod in 2008.

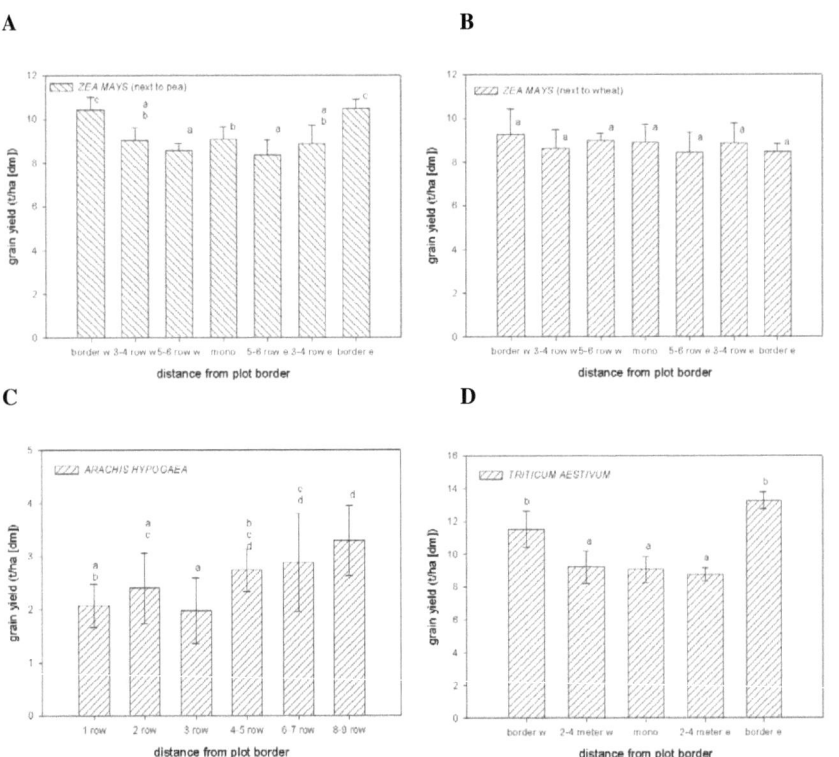

Figure 3: Intercropping as field border effect shown for selected intercropped species: (A, B) grain yield of intercropped maize, (C) peanut, and (D) winter wheat with regard to the distance from field border to monocropping in (A, B, D) southwest Germany and the (C) North China Plain in 2007/2008. Error bars indicate the standard deviation. Experimental units within each plot were border strips (0–2 m), 2–4 m distance from plot border and monocropping strips (4–6 m) for wheat in Germany. For maize in Germany and maize and peanut in China, the experimental units within each plot were two or one row, respectively. "Border w" and "border e" denote west and east plot border, respectively. Statistical analysis was done by fitting a mixed model with fixed position effects and a random effect

modeling the spatial trend by covariance structures. Positions sharing no letter are significantly different at α = 0.05 according to a t-test (Piepho, 2004)

Field trials with peanut in China: Peanut suffered from being intercropped with maize. Grain yield was significantly decreased within the first four to five strips. In 2008 and 2009, grain yield increased from strip one to strip six from 2 to 3 t [dm] ha^{-1}.

Field trials with maize in Germany and China: Maize intercropped with wheat suffered at the beginning of the growing season, because of strong competition from wheat especially for solar radiation as well as N availability. After wheat harvest, maize showed a compensation growth (Li et al., 2001) resulting in maize borderlines yielding as high as maize in monocropping. There were no significant differences between maize border strips and the strips in various distances from the border. In the year 2008, grain yield of maize in all strips were about 9 t [dm] ha^{-1} in the year 2008 and 11 t [dm] ha^{-1} in 2009.

Grain yield of maize when intercropped with pea and peanut was significantly increased within the first two to three strips in 2008. Grain yield of maize intercropped with pea declined from 10 t [dm] ha^{-1} in the first strip to 9 t [dm] ha^{-1} in monocropping.

In the year 2009, grain yield declined from 11 t [dm] ha^{-1} in the first strip to 10 t [dm] ha^{-1} in monocropping.

In the year 2008, grain yield of maize intercropped with peanut decreased from 10 t [dm] ha^{-1} in the first strip to 8 t [dm] ha^{-1} in strip seven due to an increased number of about 30 kernels per ear and an increase in TKW of about 50 g. In 2009, grain yield in the first as well as in the last measured strip of maize next to peanut was 8 t [dm] ha^{-1}.

In comparison to the year 2008, field border effects in the Chinese maize/peanut intercropping system were less distinctive in 2009, which may be attributed to the missing plant protection measures in that year.

Differences between the different distances to the border were subjected to a t-test (Fig. 3). Since the means were more than two, a multiple comparison procedure controlling the family-wise type I error rate was performed for comparison, that is, the Tukey-Kramer-test. Here, results are only briefly summarized without giving the details. Both tests led to similar results and significances, with the Tukey-Kramer-test being more conservative. The results for the wheat-maize intercropping trials in Germany were identical. The comparison between maize strip three to maize strip four when maize was grown next to pea was significant by the t-test, but not significant when using the Tukey-Kramer-test. In addition, in the Chinese peanut trials, the t-test yielded six significant differences whereas the Tukey-Kramer-test found only two.

Table 1. Akaike Information Criterion (AIC) values for different mixed models analyzed on different traits for the intercropping trial in Germany for the year 2008. Each species (maize, wheat, and pea) was analyzed separately (n.m. = not measured). Bold marked are the models with the lowest AIC value.

Germany 2008 trial	Maize (wheat)	Wheat	Maize (pea)	Pea	Maize (wheat)	Wheat	Maize (pea)	Pea	Maize (wheat)	Wheat	Maize (pea)	Pea	Maize (wheat)	Wheat	Maize (pea)	Pea
	Grain yield				TKW†				LAI				Stand density			
Baseline	176.8	112.4	158.7	113.7	223.7	57.3	165.0	120.4	45.4	19.8	52.3	n.m	56.6	44.6	68.2	88.9
Ar1/Exp without nugget	176.8	114.3	158.5	113.7	223.7	57.4	165.1	120.5	45.6	19.8	51.4	n.m	57.2	36.8	68.2	89.9
Gau without nugget	178.8	114.3	160.7	115.7	225.7	59.3	165.0	122.4	45.7	21.8	52.3	n.m	57.3	37.3	70.2	90.9
Sph without nugget	178.8	116.3	158.7	113.7	223.7	57.3	165.0	124.1	45.4	21.9	51.6	n.m	58.6	36.8	68.2	88.9
LV without nugget	189.8	114.3	169.4	118.6	239.9	65.6	181.3	122.1	47.5	19.9	49.5	n.m	62.6	**34.8**	80.8	97.5
Ar1/Exp with nugget	176.8	116.3	158.7	115.9	223.7	57.3	165.0	122.5	47.6	21.8	53.4	n.m	59.0	36.8	68.2	88.9
Gau with nugget	180.8	116.3	162.7	117.7	227.7	61.3	165.0	124.4	47.6	21.8	53.5	n.m	58.9	37.3	70.2	92.9
Sph with nugget	178.8	114.4	162.6	117.7	225.7	59.3	167.0	122.4	49.4	21.8	56.3	n.m	60.6	46.6	72.2	90.9
LV with nugget	178.8	113.8	160.6	115.6	223.7	58.0	165.1	121.9	45.5	20.3	**49.2**	n.m	57.3	36.8	71.8	89.0

† TKW: thousand kernel weight, LAI: leaf area index, Ar1: autoregressive model, Exp: exponential model, Gau: Gaussian model, Sph: spherical model, LV: linear variance (Williams, 1986).

Intercropping and Spatial Analysis

A number of spatial models were fitted for different measured traits. For the field trials in Germany, the traits grain yield, TKW, leaf area index (LAI), and stand density were analyzed for maize, pea, and wheat in the year 2008. For the year 2009 only data for the trait grain yield was available. The spatial models did not improve model performance when yield and TKW were analyzed in the year 2008 (Table 1). Also in the year 2009, the spatial models could not improve the model fit for grain yield of wheat. The baseline model had the lowest AIC value and was therefore the most suitable. By contrast, for LAI and stand density an improvement of the AIC values by using spatial models was detectable for the year 2008 as well as for grain yield of maize and pea in the year 2009 (Table 2). For these traits the linear variance model and the spherical model reached the lowest AIC values for the different species. For the trait LAI, the lowest AIC value for the linear variance model with nugget was obtained for maize next to pea in the year of 2008. For stand density in 2008, also the lowest AIC value was reached for wheat by the linear variance model without nugget (Table 1). In the year 2009, for grain yield of maize next to wheat the linear variance model with nugget showed the lowest AIC values and were therefore most suitable. The linear variance model without nugget reached the lowest AIC value for the grain yield of pea, the spherical model without nugget for the grain yield of maize next to pea (Table 2).

Two species were used in the Chinese intercropping trial, maize and peanut. A spatial analysis was done for the traits yield, TKW, and kernels/ear of maize and for peanut only the trait yield was available in the year 2008. For the year 2009, only grain yield was available for both species.

For all the traits in both years, except the grain yield of peanut in the year 2009, a spatial model had the lowest AIC value and therefore the best model fit (Tables 2 and 3). The coefficient of variation for the Chinese experiment was much higher (16.06) than the coefficient of variation for the German experiment (8.95) and therefore it can be assumed that the Chinese experiment was much more heterogeneous than the German experiment. This may explain why spatial models tended to perform better in the Chinese experiments. The extension of the baseline model by a spatial component improved the model fit for all the traits measured in China, except for grain yield of peanut in the year 2009. Different spatial models had the lowest AIC value: the Gaussian model without nugget, the AR(1) without nugget, and the linear variance model with nugget (Table 3).

Table 2. Akaike Information Criterion (AIC) values for different mixed models analyzed on different traits for the intercropping trial in Germany for the year 2009. Each species (maize, pea, and wheat) was analyzed separately. Bold marked are the models with the lowest AIC value.

Germany 2009 trial	Maize (pea) grain yield	Pea grain yield	Wheat grain yield	Maize (wheat) grain yield
Baseline	72.4	48.2	**46.6**	54.6
Ar1/Exp† without nugget	68.4	47.9	48.6	52.6
Gau without nugget	70.1	49.0	48.6	54.5
Sph without nugget	**66.4**	48.3	46.6	53.4
LV without nugget	67.7	**46.3**	50.2	51.4
Ar1/Exp with nugget	68.4	49.3	50.0	53.3
Gau with nugget	67.4	48.8	50.6	53.7
Sph with nugget	69.6	49.3	50.0	53.3
LV with nugget	69.7	47.3	48.0	**51.3**

† Ar1: autoregressive model, Exp: exponential model, Gau: Gaussian model, Sph: spherical model, LV: linear variance (Williams, 1986).

Table 3. Akaike Information Criterion (AIC) values for different mixed models analyzed on different traits for the intercropping trial in China. Each species (maize and peanut) was analyzed separately. Bold marked are the models with the lowest AIC value.

China trial	Maize 2008 grain yield	Peanut 2008 grain yield	Maize 2008 TKW†	Maize 2008 Kernels/ear	Maize 2009 grain yield	Peanut 2009 grain yield
Baseline	75.7	45.5	213.0	189.2	21.7	**61.4**
Ar1/Exp without nugget	75.5	45.1	211.5	186.5	8.7	61.4
Gau without nugget	**74.7**	45.3	**211.4**	187.1	**3.9**	61.4
Sph without nugget	74.7	45.3	213.0	186.7	6.0	61.4
LV without nugget	88.5	56.6	225.4	198.5	9.2	82.3
Ar1/Exp with nugget	77.5	43.0	213.5	188.5	7.0	61.4
Gau with nugget	76.7	43.8	213.4	189.1	5.3	65.4
Sph with nugget	77.7	43.8	215.1	191.2	8.0	63.6
LV with nugget	76.7	**42.1**	213.5	189.2	9.3	61.7

† TKW: thousand kernel weight. Ar1: autoregressive model, Exp: exponential model, Gau: Gaussian model, Sph: spherical model, LV: linear variance (Williams, 1986).

DISCUSSION
Field Border Effect sine qua non in Intercropping

Field trials in southwest Germany and in the North China Plain with different species combinations showed that if synergistic or competition effects of intercropping occur, they are mainly based on border-row effects. Significant differences in grain yield from monocropped in comparison to intercropped species were mostly restricted to the first four rows reflecting the intercropping situation and giving evidence of intercropping being associated with a field border effect (Knörzer et al., 2010).

The border effect strongly depends on the species neighboring each other. Maize next to wheat performed differently than maize next to a legume. In addition, pea performed differently than peanut even if both of them were planted next to maize. Thus, leguminous plants have to be analyzed for their intercropping suitability separately by species. Each intercropping system has its peculiarities and has to be regarded individually.

Similar studies from America (Cruse, 1996; Ghaffarzadeh et al., 1994; Iragavarapu and Randall, 1996) and China (Li et al., 2001) also indicated that intercropping effects were confined to border rows. Li et al. (2001) attributed the overall increase in yield in intercropped wheat by about 33% to inner row effects and by 67% to border row effects. Our results indicated that in wheat, the border strip effect was significant within the first 2 m or 15 rows of wheat with the first four rows mostly contributing to the observed yield increase. Studies of Cruse (1996), Ghaffarzadeh et al. (1994) and Iragavarapu and Randall (1996) had in common, that the row effects leading to an increased or decreased yield were restricted to the first two rows for the crop combination maize/legume and legume/cereal. Our results confirmed that for maize/legume the border effect was restricted to the first two rows. In contrast, maize intercropped with wheat showed no significant border effect. Legumes behaved differently. Pea intercropped with maize in the German experiment showed no significant border row effect. In the Chinese experiment, yield of peanut when intercropped with maize was affected by its neighboring plants up to row six.

Indeed, in other publications the row configuration had significant effects on intercropping system performance (Chen et al., 2004; Mandal et al., 1996; Meena et al., 2006; Rao and Willey, 1983), depending on the number of rows or strip width of each component crop, but the effects were restricted to border row effects. Maize combined with legumes is influenced within the first two rows. More than two rows of cereals like oat, wheat, or barley (*Hordeum vulgare* L.) profit from intercropping with legumes or maize (Banik et al., 2006; Haymes and Lee, 1999; Gunes et al., 2007; Ghaley et al., 2005; Song et al., 2007), but the border effect of a strip could be restricted to the first meter. Legumes performance is inconsistent, depending on crop component and leguminous species.

So far, intercropping has mostly been considered as a secluded cropping system within one field. But in African and Asian countries, where intercropping is widespread, the system can be extended to a much larger scale: common on-field intercropping goes along with small field size on average, low mechanization level and, hence, small field border distances. For example, in China, where the average farm size is around 0.1 ha, small fields alternate as strips with different crops grown on them, thus turning field boundaries into a kind of unplanned intercropping at a larger scale (Knörzer et al., 2009). Hence, the more fragmented the agricultural landscape is, the more relevant the borders can become.

If we consider crop performance with special regard to grain yield in those fragmented agricultural landscapes, it might not be sufficient to analyze or gauge the yield of a small field as a whole as there is a yield distribution or range within those fields. The smaller the field size and the higher the proportion of field borders, the more relevant are yield increases or decreases due to border effects. Cultivation practice or variety influence could be overestimated when neglecting competition effects. Thus, a spatial analysis of those small fields could help to take these factors into account.

Intercropping and Spatial Analysis

In comparison to other intercropping studies, the ANOVA was done with respect to spatial distribution of experimental plots taking into account that intercropped species can rarely be randomized. In general, intercropping has been analyzed assuming that the combination of species and the position of rows of each crop within the intercropping system is the experimental treatment (Agyare et al., 2006; Chen et al., 2004). Studies by Ghaffarzadeh et al. (1994), Cruse (1996), and Iragavarapu and Randall (1996) analyzed each crop separately, taking the row position within the strips as treatment factor, similar to our study.

Ghaffarzadeh et al. (1994) realized the problem of restricted randomization within intercropping systems, pointing out that their strip-position analysis differed from the customary replicated single-factor design in that the treatment could not be assigned a random position within the block. In addition, Connolly et al. (2001) found in their review that the four most common designs in intercropping experiments are the simple pairwise design, which consists of a single mixture perhaps repeated under a range of levels of a treatment, the additive and replacement series design, and the response model design. Response model designs and some additive series experiments required regression methods, while bivariate methods were used in simple pairwise designs. But mostly, the multivariate nature of intercropping (Connolly et al., 2001; Federer, 1998) as well as the spatial trend within nonrandomized strip intercropping experiments were not taken into account. To the best of our knowledge, the latter was taken into account in this study for the first time. Gomes et al. (2008) proposed another statistical approach analyzing a randomized complete block and a split-plot design within an ANOVA framework. Again, the common assumptions were made defining the treatments as the crop combination and the sole crops, but the analysis of variance was done by applying the normal theory or nonparametric methods to the efficiency measurements, creating efficiency ranks. A key assumption of their analysis was a random allocation of treatments to the experimental plots (Gomes et al., 2008), which is not justified in strip intercropping designs.

The problems in analyzing intercropping performance statistically were already discerned by Connolly et al. (2001), Federer (1993, 1998), and Gomes et al. (2008). In our study different spatial components were added to the baseline model to account for spatial trend, to analyze intercropping performance, to assess as border effects, and to check whether the baseline model is suitable for analyzing intercropping experiments or if spatial models have to be used.

In the German experiment, yield and some yield components could be adequately analyzed with the baseline model based on the AIC values. No further improvement by using spatial methods was reached for the traits yield and TKWs in 2008. It is assumed that this is caused by the more homogenous German experiment as indicated by a lower coefficient of variation in comparison to the Chinese experiment in 2008.

By contrast, for LAI and stand density an improvement of the AIC values by using spatial models was detectable in 2008. Also an improvement by spatial models was found for grain yield of the different species in 2009.

However the numerical differences between AIC values of different models were usually not very big, so the model selection was certainly not perfect.

As a general strategy, we propose to start the analysis of intercropping experiments by a baseline model and to extend by spatial add-on components, if the model fit can be improved (Piepho et al., 2008; Piepho and Williams, 2010; Müller et al., 2010).

ACKNOWLEDGMENTS

The authors' research topic is embedded in the International Research Training Group of the University of Hohenheim and China Agricultural University, entitled "Modeling Material Flows and Production Systems for Sustainable Resource Use in the North China Plain". We thank the German Research Foundation (DFG) and the Ministry of Education (MOE) of the People's Republic of China for financial support.

REFERENCES

Agyare, W.A., V.A. Clottey, H. Mercer-Quarshie, and J.M. Kambiok. 2006. Maize yield response in a long-term rotation and intercropping systems in the Guinea savannah zone of Northern Ghana. J. Agron. 5:232–238.

Bailey, R.A., J.M. Azais, and H. Monod. 1995. Are neighbour methods preferable to analysis of variance for completely systematic designs? Silly designs are silly! Biometrika 82:655–659.

Banik, P., A. Midya, B.K. Sarkar, and S.S. Ghose. 2006. Wheat and chickpea intercropping systems in an additive series experiment: Advantages and weed smothering. Eur. J. Agron. 24:325–332.

Bartlett, M.S. 1978. Nearest neighbor models in the analysis of field experiments (with Discussion). J. R. Stat. Soc. Ser. B 40:147–174.

Beets, W.C. 1982. Multiple cropping and tropical farming systems. Westview Press, Boulder, CO.

Chen, C., M. Westcott, K. Neill, D. Wichmann, and M. Knox. 2004. Row configuration and nitrogen application for barley-pea intercropping in Montana. Agron. J. 96:1730–1738.

Cochran, W.G., and G.M. Cox. 1957. Experimental designs. John Wiley & Sons, New York.

Cohen, R. (ed.). 1988. Satisfying Africa's food needs: Food production and commercialization in African agriculture. Book series: Carter Studies on Africa. Rienner, Boulder, CO.

Connolly, J., H.C. Goma, and K. Rahim. 2001. The information content of indicators in intercropping research. Agric. Ecosyst. Environ. 87:191–207.

Cruse, R.M. 1996. Strip intercropping: A CRP conversion option. Conservation Reserve Program: Issues and options. CRP-17. Univ. Ext., Iowa State Univ., Ames.

Cullis, B.R., and A.C. Gleeson. 1991. Spatial analysis of field experiments - An extension to two dimensions. Biometrics 47:1449–1460.

Edmondson, R.N. 2005. Past developments and future opportunities in the design and analysis of crop experiments. J. Agric. Sci. 143:27–33.

FAO (ed.). 2006. The state of food and agriculture. FAO Agric. Ser. 27. FAO, Rome.

Federer, W.T. 1993. Statistical design and analysis for intercropping experiments, Volume I: Two crops. Springer, Berlin.

Federer, W.T. 1998. Statistical design and analysis for intercropping experiments, Volume II: Three or more crops. Springer, Berlin.

Fernandez, C., F.J. Acosta, G. Abella, F. Lopez, and M. Diaz. 2002. Complex edge effect fields as additive processes in patches of ecological systems. Ecol. Modell. 149:273–283.

Fisher, R.A. 1925. Statistical methods for research workers. 1st ed. Oliver and Boyd, Edinburgh.

Ghaffarzadeh, M., F.G. Prechac, and R.M. Cruse. 1994. Grain yield response of corn, soybean, and oat grown in a strip intercropping system. Am. J. Altern. Agric. 9:171–177.

Ghaley, B.S., H. Hauggaard-Nielsen, H. Hogh-Jensen, and E.S. Jensen. 2005. Intercropping of wheat and pea as influenced by nitrogen fertilization. Nutr. Cycling Agroecosyst. 73:201–212.

Gilmour, A.R., B.R. Cullis, and A.P. Verbyla. 1997. Accounting for natural and extraneous variation in the analysis of field experiments. J. Agric. Biol. Environ. Stat. 2:269–293.

Gleeson, A.C. 1997. Spatial analysis. p. 68–85. *In* R. Kempton and P.N. Fox (ed.) Statistical methods for plant variety evaluation. Chapman & Hall, London.

Gleeson, A.C., and B.R. Cullis. 1987. Residual maximum likelihood (REML) estimation of a neighbour model for field experiments. Biometrics 43:277–288.

Gomes, E.G., G. D. S. Souza, and L.J. Vivaldi. 2008. Two-stage interference in experimental design using DEA: An application to intercropping and evidence from randomization theory. Pesquisa Operacional 28:339–354.

Gunes, A., A. Inal, M.S. Adak, M. Alpaslan, E.G. Bagci, T. Erol, and D.J. Pilbeam. 2007. Mineral nutrition of wheat, chickpea and lentil as affected by mixed cropping and soil moisture. Nutr. Cycling Agroecosyst. 78:83–96.

Haymes, R., and H.C. Lee. 1999. Competition between autumn and spring planted grain intercrops of wheat (*Triticum aestivum*) and field bean (*Vicia faba*). Field Crops Res. 62:167–176.

Iragavarapu, T.K., and G.W. Randall. 1996. Border effects on yields in a strip intercropped soybean, corn, and wheat production system. J. Prod. Agric. 9:101–107.

John, J.A., and E.R. Williams. 1995. Cyclic and computer generated designs. 2nd ed. Chapman and Hall, London.

Knörzer, H., S. Graeff -Hönninger, B. Guo, P. Wang, and W. Claupein. 2009. The rediscovery of intercropping in China: A traditional cropping system for future Chinese agriculture. *In* E. Lichtfouse (ed.) Springer Series: Sustainable agriculture reviews 2: Climate change, intercropping, pest control and benefi cial microorganisms. Springer Science + Business Media, Berlin.

Knörzer, H., S. Graeff -Hönninger, B.U. Müller, H.-P. Piepho, and W. Claupein. 2010. A modeling approach to simulate eff ects of intercropping and interspecific competition in arable crops. Int. J. Information Syst. Social Change Special Issue (in press).

Li, W. 2001. Agro-ecological farming systems in China. *In* J.N.R. Jeffers (ed.) Man and the biosphere series, Vol. 26. Taylor & Francis, Paris.

Li, L., S.M. Li, J.H. Sun, L.L. Zhou, X.G. Bao, H.G. Zhang, and F. Zhang. 2007. Diversity enhances agricultural productivity via rhizosphere phosphorus facilitation on phosphorus-deficient soils. Proc. Natl. Acad. Sci. USA 104:11192–11196.

Li, L., J. Sun, F. Zhang, X. Li, S. Yang, and Z. Rengel. 2001. Wheat/maize or wheat/soybean strip intercropping. I. Yield advantage and interspecific interactions on nutrients. Field Crops Res. 71:123–137.

Mandal, B.K., D. Das, A. Saha, and Md. Mohasin. 1996. Yield advantage of wheat (*Triticum aestivum*) and chickpea (*Cicer arietinum*) under different spatial arrangements in intercropping. Indian J. Agron. 41:17–21.

Meena, O.P., B.L. Gaur, and P. Singh. 2006. Effects of row ration and fertility levels on productivity, economics and nutrient uptake in maize (*Zea mays*) + soybean (*Glycine max*) intercropping system. Indian J. Agron. 51:178–182.

Müller, B.U., K. Kleinknecht, J. Möhring, and H.-P. Piepho. 2010. Comparison of spatial models for sugar beet and barley trials. Crop Sci. 50:May-June.

Piepho, H.-P. 2004. An algorithm for a letter-based representation of all-pairwise comparisons. J. Comput. Graph. Statist. 13:456–466.

Piepho, H.-P., A. Büchse, and C. Richter. 2004. A mixed modelling approach for randomized experiments with repeated measures. J. Agron. Crop Sci. 190:230–247.

Piepho, H.-P., C. Richter, and E.R. Williams. 2008. Nearest neighbour adjustment and linear variance models in plant breeding trials. Biometrical J. 50:164–189.

Piepho, H.-P., and E.R. Williams. 2010. Two-dimensional linear variance structures for plant breeding trials. Plant Breed. 129:1–8.

Rao, M.R., and R.W. Willey. 1983. Effects of pigeonpea plant population and row arrangement in sorghum/pigeonpea intercropping. Field Crops Res. 7:203–212.

SAS Institute. 2009. The SAS system for Windows. Release 9.2. SAS Inst., Cary, NC.

Schabenberger, O., and C. Gotway. 2005. Statistical methods for spatial data analysis. CRC Press, Boca Raton, FL.

Schwarzbach, E. 1984. A new approach in the evaluation of field trials. The determination of the most likely genetic ranking of varieties. Vortr. Pflanzenzuecht. 6:249–259.

Song, Y.N., F.S. Zhang, P. Marschner, F.L. Fan, H.M. Gao, X.G. Bao, J.H. Sun, and L. Li. 2007. Effect of intercropping on crop yield and chemical and microbiological properties in rhizosphere of wheat (*Triticum aestivum* L.), maize (*Zea mays* L.), and faba bean (*Vicia faba* L.). Biol. Fertil. Soils 43:565–574.

Vandermeer, J. 1989. The ecology of intercropping. Cambridge Univ. Press, Cambridge.

Wilkinson, G.N., S.R. Eckert, T.W. Hancock, and O. Mayo. 1983. Nearest neighbour analysis of field experiments (with discussion). J. R. Stat. Soc. Ser. B 45:151–211.

Williams, E.R. 1986. A neighbor model for field experiments. Biometrika 73:279–287.

Wolfinger, R.D. 1996. Heterogeneous variance-covariance structures for repeated measures. J. Agric. Biol. Environ. Statistics 1:205–230.

Wubs, A.M., L. Bastiaans, and P.S. Bindraban. 2005. Input levels and intercropping productivity: Exploration by simulation. Note 369. Plant Res. Int., Wageningen, the Netherlands.

7 Chapter IV / Excursus:

Model-based approach to quantify and regionalize peanut production in the major peanut production provinces in the People's Republic of China

---

PUBLICATION IV:

Knörzer, H., Graeff-Hönninger, S., and Claupein, W. (2010): Model-based approach to quantify and regionalize peanut production in the major peanut production provinces in the People's Republic of China. GI-Edition - Lecture Notes in Informatics „Precision Agriculture Reloaded – Informationsgestützte Landwirtschaft", pp. 101-104.

Chapter IV is a potential peanut yield study in four major peanut producing provinces in China without differentiating between intercropping and monocropping systems. Therefore, the study does not predominantly deal with intercropping and could be classified in this context as an excursus. As the peanut crop was part of the Chinese field experiments, the potential yield study was included into this thesis in order to gain an overview over yield and yield potential in the provinces Shandong, Anhui, Hebei, and Henan, the latter three being part of the study region 'North China Plain'. Like in chapter III, the DSSAT model was chosen. Peanut yield data for the different provinces was taken from the China Provincial Statistical Yearbooks, and collected soil as well as weather data from previous project members could be used to evaluate and validate the model, to account for provincial differences, and to analyze yield potential in a long-term point of view. Special regard was given to water supply and water shortage as precipitation is a major limiting factor of peanut production within the North China Plain.

Model-based approach to quantify and regionalize peanut production in the major peanut production provinces in the People's Republic of China

Heike Knörzer, Simone Graeff-Hönninger, Wilhelm Claupein
Institut für Pflanzenbau und Grünland
Universität Hohenheim
Fruwirthstrasse 23
70593 Stuttgart

Article from the GI-Edition - Lecture Notes in Informatics „Precision Agriculture Realoaded – Informationsgestützte Landwirtschaft" (2010), pp. 101-104, with permission, copyright © 2010, Gesellschaft für Informatik e.V. (GI).

ABSTRACT

China is the largest peanut producer in the world and peanut therefore an essential economic product earning significant income for China's farmers. Major provinces for peanut production are located in the middle and eastern provinces. In these regions, drought stress between germination and pod setting, could be severe and yield decline because of uneven rainfall and climate variability. Four provinces were selected for modeling and simulating large area yield estimation in order to evaluate potential yield with respect to average rainfall in the individual regions. Measured, average yield during the evaluated years ranged from 2918 kg ha^{-1} to 3969 kg ha^{-1} with a mean yield of 3420 kg ha^{-1} and a mean modeled yield of 3422 kg ha^{-1}. The model showed a good fit between observed and simulated data with a RMSE of 252. Model error was 7.4 %.

INTRODUCTION

Concerning peanut and peanut production, China speaks in superlatives. With more than 290 000 t shelled and unshelled peanuts in 2007, the country is the largest peanut exporter in the world according to the FAO with one third of the world market share [Ch09]. China has one fifths of the world area under peanut and it produces more than two fifths of the total world peanut production [Ga04]. In addition, peanut is the major oilseed crop in China making 40 % of the land's total oilseed production and 25 % of the cropped area [Ga96]. Hence, peanut is an essential economic product for the country. Since the foundation of the PR China, more than 200 varieties have been bred and more than 80 have been introduced and used [Ga96] showing the importance of the crop. Approximately 70 % of the production takes place in five provinces [Ga04] with Shandong, Henan and Hebei on top. Seven agro ecological zones are determined with peanut production predominant in the Chinese middle and eastern provinces. Highest yields and the largest area under peanut production are achieved in Shandong and Henan province, both part of the North China Plain.

Generally one crop or three crops per two years are grown in these regions with peanut usually being intercropped with wheat, maize, bean or sweet potato. Since the 1980's, when improved management practices like the polythene mulching and improved varieties have been introduced, peanut yield increased steadily [Ga04] [Ga96].

Nevertheless, there are constraints about substantial yield gaps [JN96] between yields realized by farmers and those recorded from research stations or potential yield estimations. Yield fluctuates strongly due to climate variation and uneven adoption of improved technology [Ga04]. The objectives of the study were to quantify the production potential of peanut in Anhui, Hebei, Henan and Shandong provinces with special regard to water demand during the growing season. Water deficit is a major constrain in peanut production [Ri08], especially during the critical period of pod set which results in reduced pegging. An expert discussion at an international workshop about situation and prospects for groundnut production in China (1996) ranked drought stress on top of the major constraints for peanut production, ahead of acid soils, pests and diseases and cold temperature. For that purpose, the Decision Support System for Agrotechnology Transfer (DSSAT) [Jo03] crop growth model Vs. 4.5 was evaluated and validated using five years of yield data from the different provinces. Scenarios were driven using average weather data from 1976-2005 and eleven meteorological stations across the North China Plain.

MATERIALS AND METHODS

Within the DSSAT crop growth model, CROPGRO-Peanut is a generic grain legume model that computes crop growth processes including phenology, photosynthesis, plant nitrogen, carbon demand, and growth partitioning. In addition, the plant development and growth module is linked to soil-plant-atmosphere modules. Hence, the model has the potential for large area yield estimation by input of soil and daily weather data [Ga06]. For evaluation and validation of the CROPGRO-Peanut model, weather data from local weather stations of each province was taken [Bi08] [Ch07]. Physical properties of the provincial predominant soil texture classes [Xi86] silt (Anhui, Hebei, Henan) and sandy loam (Shandong) were derived from [Ch97], [Bö04], and [Bi08]. Average yield from Anhui, Hebei, Henan and Shandong provinces were taken using data from [Ch01]. A cross validation was done for Anhui for the years 2001/02/04/05, for Shandong for the years 2001/02/03/04, for Hebei for the years 2001/02/03/04/05 and for Henan for the years 2002/04/05 using the pre-evaluated DSSAT cultivar 'Chinese TMV2 TAM' (Nongshen type) for Anhui, Hebei and Henan and the 'Florigiant new' (Virginia type) for Shandong as initial point. According to the cross validation, the cultivars' coefficients were slightly modified for the four provinces. An average plant density of 40 plants m^{-1} within a hill planting system was used, taking the polythene mulching practice into account. As peanut is a leguminous plant, a balanced fertilization of 20 kg N ha^{-1} at sowing was applied for model evaluation. For each province, yield potential was simulated for average daily incoming solar radiation, minimum temperature, maximum temperature and rainfall using eleven meteorological weather stations' data. Subsequently, irrigated and rainfed scenarios within a sensitivity analysis were driven in order to detect whether long-term average rainfall could meet the water demand.

RESULTS AND DISCUSSION

The model showed a good fit between observed and simulated yield after the cross evaluation and validation procedure (Fig. 1). Mean observed yield of Anhui was 3625 kg ha^{-1}, mean simulated yield 3640 kg ha^{-1} with a RMSE of 300 kg ha^{-1} (model error = 8.3 %). Mean observed yield for Hebei was 3028 kg ha^{-1}, mean simulated yield 3030 with a RMSE of 315 kg ha^{-1} (model error = 10.4 %). For Henan, mean observed yield was 3410 kg ha^{-1} and mean simulated yield 3391 with a RMSE of 46 kg ha^{-1} (model error = 1.4 %). Mean observed yield in Shandong was 3713 kg ha^{-1}, mean simulated yield 3718 kg ha^{-1} with a RMSE of 196 kg ha^{-1} (model error = 5.3 %). Overall cross validation had a RMSE of 252 kg ha^{-1} (model error = 7.37 %). Mean observed yield of all provinces was 3420 kg ha^{-1}, mean simulated yield 3422 kg ha^{-1}. For each province, three simulation scenarios were chosen: 1.) 20 kg N ha^{-1} at sowing with supplement irrigation; 2.) 20 kg N ha^{-1} at sowing, rainfed; 3.) neither nitrogen nor water stress. The results are shown in Table 1.

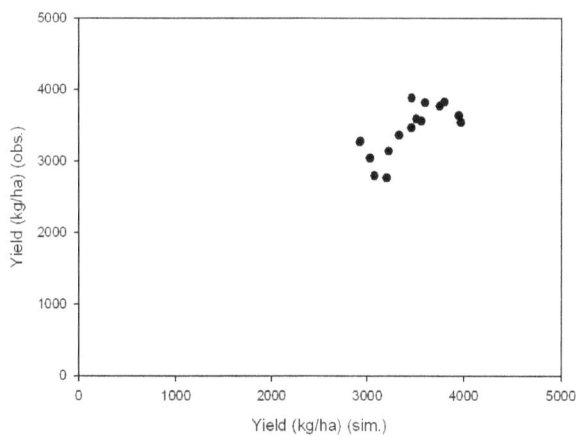

Figure 1: Mean observed peanut yield of Anhui, Hebei, Henan and Shandong provinces versus mean simulated yield after cross evaluation and validation of the CROPGRO-Peanut model

Table 1: Simulated peanut yield of Anhui, Hebei, Henan and Shandong provinces using average weather data from 1976-2005 and three different scenarios

Scenario	Anhui Yield (kg ha^{-1})	Hebei Yield (kg ha^{-1})	Henan Yield (kg ha^{-1})	Shandong Yield (kg ha^{-1})
20 kg N ha^{-1}, irrigated	3848	3216	3877	4485
20 kg N ha^{-1}, rainfed	3848	3254	3914	2746
no N/water stress	3903	3244	3927	4482

As reported, peanut is very susceptible for water deficit. In the North China Plain, 50-75 % of the total rainfall occurs between July and September. In most years, the amount of rainfall is enough to satisfy the demand. Water shortage may occur at the beginning of the growing season of peanut as peanut is generally sown between mid of April and mid of May in these regions, and may last until the critical phase of pod setting around 50 days after sowing. The simulation showed that for Anhui and Henan, the long-term average rainfall met the demand of peanut. No supplement irrigation would be needed.

In contrast, the average rainfall in Hebei and especially in Shandong was not enough to satisfy the demand. Without irrigation, peanut yield decreased notable in Shandong. Shandong is the most important peanut region in China [Ch09] [Ga04] with the highest yield potential. The sandy soils predominant in Shandong were more likely to desiccate than the silty soils in the other provinces. [Ri08] reported from Argentina, that water stress promoted a significant decline up to 73 % in peanut seed yield because of reduced seed and pod numbers. Accordingly, a notable reduction in pod number for Shandong was simulated by the model. The CROPGRO-Peanut model was able to simulate and estimate large area yield for the four major peanut producing provinces in China. To study the potential yield of peanut with special regard to water demand, the approach was useful comparing possible lacks in demand with average rainfall. Further on, the model setting could be used for additional scenario and sensitivity analysis[1].

---

[1] We thank the German Research Foundation (DFG) and the Ministry of Education (MOE) of the People's Republic of China for financial support (GRK 1070).

REFERENCES

[Bi08] Binder, J. et al.: Model approach to quantify production potentials of summer maize and spring maize in the North China Plain. In: Agronomy Journal 100, 2008, p. 862-873.

[Bö04] Böning-Zilkens, M.I: Comparative appraisal of different agronomic strategies in a winter wheat-summer maize double cropping system in the North China Plain with regard to their contribution to sustainability. Diss. Universität Hohenheim, Aachen, 2004.

[Ch09] Chen, C. et al.: Competitiveness of peanuts: United States versus China. The University of Georgia, Cooperative Extension, Research Bulletin 430, 2009.

[Ch07] China Meteorological Administration: Weather data. Beijing, 2007.

[Ch01] China Provincial Statistical Yearbooks, All China Data Center, University of Michigan, 2001-2005.

[Ch97] Chinese Academy of Science: Chinese soil survey 1997. Beijing, 1997.

[Ga96] Gai, S. et al.: Present situation and prospects for groundnut production in China. In: Achieving high groundnut yields (Gowda, C.L.L., Nigam, S.N., Johansen, C. and Renard, C., Ed.), India, 1996, p. 17-26.

[Ga04] Gang, Y.: Peanut production and utilization in the People's Republic of China. In: Peanut in local and global food systems series report no. 4 (Rhoades, R.E., Ed.), University of Georgia, 2004.

[Ga06] Garcia, G.Y. et al.: Analysis of the inter-annual variation of peanut yield in Georgia using a dynamic crop simulation model. In: Transactions of the ASABE 49, 2006, p. 2005-2015.

[JN06] Johansen, C., Nageswara Rao, R.C.: Maximizing groundnut yield. In: Achieving high groundnut yields (Gowda, C.L.L., Nigam, S.N., Johansen, C. and Renard, C., Ed.), India, 1996, p. 117-127.

[Jo03] Jones, J.W. et al.: The DSSAT cropping system model. In: European Journal of Agronomy 18, 2003, p. 235-265.

[Ri08] Ricardo, J.H. et al.: Seed yield determination of peanut crops under water deficit: soil strength effects on pod set, the source-sink ratio and radiation use efficiency. In: Field Crops Research 109, 2008, p 24-33.

[Xi89] Xi, Y. (Ed.): Atlas of the People's Republic of China. 1st ed., Foreign Language Press, China Cartographic Pub. House, Beijing, 1989.

8 Chapter V:
Integrating a simple intercropping algorithm into CERES-wheat and CERES-maize with particular regard to a changing microclimate within a relay-intercropping system

PUBLICATION V:
Knörzer, H., Grözinger, H., Graeff-Hönninger, S., Hartung, K., Piepho, H.-P., and Claupein, W. (2011): Integrating a simple intercropping algorithm into CERES-wheat and CERES-maize with particular regard to a changing microclimate within a relay-intercropping system. Field Crops Research 121 (2), pp. 274–285.

The modeling approach presented in chapter II was further developed and is presented in the following chapter. In comparison to the one year dataset used in chapter II, the complete dataset of the field experiments from 2007 to 2009 was taken. In addition, the influence of different or changing microclimates within two different intercropping systems were analyzed and considered in more detail. Two major aspects contributed to the outcome of chapter V. The first one was for sure that the first model run (chapter II) and the concept of introducing a shading algorithm to modify the weather input for the model instead of introducing an additional submodel routine into DSSAT seemed to be promising. Thus, an additional dataset was necessary to test the model performance further on. The second one was that there was some kind of development within the ongoing field trials between the first and the second experimental year. During and after data collection and analysis in 2008, different aspects, effects, and issues evolved. Competition within intercropping seemed not to be restricted to competition for solar radiation solely, but in contrast much more influenced by multiple factors of changing microclimate. Thus, spatial and temporal measurements like wind speed and soil temperature had to be extended in the second year in comparison to the first year. Additional N_{min} samples were needed, too. As a result, the shading algorithm as well as microclimate influences could be considered in the modeling approach and the simulation of strip intercropping and were the main objectives of the following chapter.

8. Chapter V

Integrating a simple shading algorithm into CERES-wheat and CERES-maize with particular regard to a changing microclimate within a relay-intercropping system

H. Knörzer[a], H. Grözinger[b], S. Graeff-Hönninger[a], K. Hartung[c], H.-P. Piepho[c], W. Claupein[a]

[a] University of Hohenheim, Institute of Crop Science, Fruwirthstraße 23, 70593 Stuttgart, Germany
[b] University of Hohenheim, Experimental Station 'Ihinger Hof', 71272 Renningen, Germany
[c] University of Hohenheim, Bioinformatics, Fruwirthstraße 23, 70593 Stuttgart, Germany

Article from Field Crops Research 121 (2011): 274–285, with permission, copyright © 2010 Elsevier B.V..

ABSTRACT

Wheat/maize related multi-cropping systems are the dominant cropping systems in North China. To improve and adjust those systems, and to study competition effects within intercropping, extended field experiments are necessary. As field experiments are time consuming, laborious and expensive, a viable alternative is the use of crop growth models that can quantify the effects of management practices on crop growth and productivity. Field experiments showed that intercropped maize yielded as high as monocropped maize, and grain yield of intercropped wheat increased by up to 32%. Based on a process-oriented modeling approach, this study focuses on analyzing and modeling competitive relationships in a wheat/maize relay intercropping system with regard to yield, solar radiation and microclimate effects. A simple shading algorithm was applied and integrated into the CERES models, which are part of the DSSAT software shell vs. 4.5. The algorithm developed estimates the proportion of shading as affected by neighbouring plant height. The model was tested to investigate the applicability of this shading algorithm within the CERES models in the simulation of grain yield and dry matter yield of wheat and maize. Model error of grain and dry matter yield for both species was below 10%. There was a tendency for grain yield to be simulated adequately but for dry matter yield to be slightly underestimated. Increased top soil temperature in intercropped wheat increased the mineralization of nitrogen and improved N supply. The wheat/maize system was N efficient. Thus, N dynamics were taken into account for simulation as well as CO_2 dynamics based upon modified wind speed. Wheat border rows were exposed to a higher wind speed until mid-June and to reduced wind speed thereafter. As a result, solar radiation, soil temperature and wind speed differed between monocropping and intercropping and could provide a starting point for simulating intercropping. Microclimate effects are often small, subtle or non-existent, while spatial and climate variability and the heterogeneity of plant populations can be considerable. Quantifying microclimatic effects may prove difficult but should not be neglected when simulating intercropping systems.

INTRODUCTION

Smallholder farming, tradition, scarcity of land, sustainability, increased resource use efficiency and yield stability, taken together, lead to a great diversity of intercropping systems all over the world, especially in Africa and Asia. In times of dwindling food security worldwide, the promotion of local food production and ecological constrains of production are key issues. Intercropping, defined as the growing of two or more crops simultaneously on the same area, is an increasingly relevant topic. As the great variety of species combinations in intercropping systems make field experiments time consuming and expensive, modeling of intercropping is an attractive alternative. Various models have attempted to deal with intercropping and interspecific competition (Caldwell, 1995; Keating and Carberry, 1993; Knörzer et al., 2010a). Most studies investigated and modeled cereal/legume intercropping systems as these are common and widespread, and the systems bear the potential for more sustainable farming. The combination of a smaller legume and a taller cereal differing in morphology, phenology and physiology is ideal for research purposes, especially when studying competition for nitrogen.

In China intercropping is a common practice (Knörzer et al., 2009). In particular, wheat/maize relay multi-cropping systems are widespread in the North China Plain and the Yellow and Huai River Valleys (Meng et al., 2006; Wang et al., 2009). Wheat/maize relay intercropping systems consist of a few rows of winter wheat (*Triticum aestivum* L.) sown with usual row spacing and with space left between each wheat strip for a later maize (*Zea mays* L.) sowing. Before China focused on breeding early maturing wheat and maize varieties, relay intercropping of wheat and maize was the only option of realizing two harvests in one year while expanding the growing season for maize. Moreover, this system could be mechanized to a certain extent. Nowadays, the relay intercropping area is declining in the North China Plain, but the practice is still common in the northern and north-western parts of the country where the growing season is short.

Although a few models or intercropping submodels are quite advanced in simulating interspecific competition for solar radiation or nitrogen, amore comprehensive modeling of intercropping comprising various species remains a complex problem to solve. The modeling approaches used to date do not go beyond modeling the diverse competition effects. They merely simulate traits such as grain yield from field experiments adequately but without providing knowledge gain throughout the modeling process. Knowledge gain in this context means that the modeling process helps to study and explain underlying physiological processes for plant growth in more detail than with measured data. This may lead to further improved scientific theories. Through the modeling process, conclusions for adjusting and improving cropping systems may also be drawn. Hence, the full use of the value-added chain of a crop growth model offers the possibility of increasing the knowledge about plant growth including plant–soil, plant–atmosphere and plant–management interactions more comprehensively. It is important to develop, evaluate and validate those models intensively and carefully, but modeling and simulation should aim to improve intercropping analysis and hence, to optimize intercropping systems in the field, for example, by finding the optimum number of crop rows, optimized fertilization and species combination.

Based on a process-oriented modeling approach, this study emphasizes analyzing and modeling competitive relationships within a wheat/maize relay intercropping system with regard to yield, solar radiation and synergistic or shelter effects of microclimate (Harris and Natarajan, 1987; Meinke et al., 2002). Currently models distributed in the DSSAT (Decision Support System for Agrotechnology Transfer) modeling package (Jones et al., 2003) do not account for the effects of competition in mixed cropping systems. A basic description of the relevant indicator parameters will be given and the transfer of a simple shading algorithm into the DSSAT CERES-wheat and CERES-maize models will be shown. In addition, a basic methodology for using a process-oriented model for simulating an intercropping scenario will be generated and tested based on a two-year dataset.

MATERIALS AND METHODS

Field experiments

Field trials were conducted in southwest Germany (48.46°N and 8.56°E) at the University of Hohenheim experimental station 'Ihinger Hof' during the growing seasons 2007/08 (2008) and 2008/09 (2009). At this location, the average rainfall per year is around 690 mm with an average temperature of 7.9°C. The soils of the Ihinger Hof site, located in the climatic region "Upper Neckar-Valley" are mainly medium-deep para-brown colored soils over loess. The drainage is moderately well and the stone percentage is about 1%. The soils, classified as Orthic Luvisols (IUSS Working Group WRB, 2007), are silty loamy with clay enriched underside. Clay percentage is about 30%.

In the year 2008, the rainfall between March and October was 683 mm, and 598 mm in the year 2009. Continuous measurements of soil water content with a Trime-TDR-system (time domain reflectometry) from IMKO GmbH (Ettlingen/Germany) indicated that no water shortage occurred during the two growing seasons.

The experiments comprised a maize/wheat relay intercropping system (termed 'intercropping' in the following) within a non-randomized complete block design and four replications. The scope for randomization was restricted as strip intercropping trials necessitate alternating strips (Knörzer et al., 2010b). Each replication consisted of two plots. The two species were planted in an alternate pattern. Each plot was 10 m ×10 m for wheat, and 12 m × 10 m for maize, and included five subplots (5 × 2 m × 10 m) for wheat, and eight subplots (8 × 1.5 m × 10 m) for maize. Within those subplots, data was collected in order to detect crop performance differences between different distances from the plot border. The plots were large enough to reflect monocropping within the central subplot. Row orientation was from north to south.

The experimental arrangement did not fully reflect a relay intercropping system but was chosen in order to detect differences in crop performance as dependent on distance from plot border or neighbouring species. If the appropriate distance from a neighbouring species is unknown or just assumed, the determination of 'influenced' and 'uninfluenced' plant row is rather vague. Thus to guarantee the distinction between monocropping and intercropping, the plots should be as near as possible to ensure similar environmental circumstances, and as large as possible to clearly identify influenced and uninfluenced rows.

The wheat variety 'Cubus' was sown in October 2007 and 2008 with a row spacing of 13 cm and a plant density of 300 plants per m^2. Maize cv. 'Companero' was sown in May 2008 and 2009 with a row spacing of 75 cm and a plant density of 10 plants per m^2.

Wheat was fertilized with 160 kg N ha^{-1}, split into three applications (60/60/40) of Nitro-chalk. Maize was fertilized once with 160 kg N ha^{-1} (ENTEC). Plant protection was carried out according to 'Good Agricultural Practice'. During both growing seasons neither water nor nitrogen (N) stress occurred, so differences in plant growth and yield performance could be attributed to intra- and interspecific competition.

Three sequential harvests according to the DSSAT cropping system model (DSSAT-CSM) guide (Hoogenboom et al., 1999) were carried out as square meter cuts, and grain yield, dry matter and nitrogen concentration of plants were analyzed. In addition, the mineralized soil nitrogen content (N_{min}) was determined at the beginning of the growing season and after harvest. In addition, five N_{min} samples were taken weekly between May and June in wheat and two in maize, in order to detect differences between N budgets among subplots. Soil temperature was determined in 5 cm soil depth with the testo 925 (Testo AG, Lenzkirch/Germany) on a weekly basis. Solar radiation was also determined weekly with the AccuPAR LP-80 (UMS, München/Germany) and the testo 545 (Testo AG, Lenzkirch/Germany). Solar radiation within the field plots was measured at the top of the canopies. Additionally, the local weather station measured daily total radiation. Wind speed was measured at the top of the canopies with a cup anemometer 'compact' (Thies Clima, Göttingen/Germany). As such, measurement heights of solar radiation and wind speed on top of the canopies changed during the growing season due to plant growth. Growing stages and plant height were recorded on a weekly basis. After the final harvest, yield and yield components such as thousand kernel weight (TKW), tiller number, ears per plant as well as N concentration and N uptake were determined. Wheat was harvested at the end of July from a 1.0 m² area. Maize was harvested in mid- October with a harvester.

Statistical analysis and model testing

Statistical analysis to detect differences in dry matter and grain yield between subplots within a plot with different distances from the plot border was carried out separately for each species in the trial. Analysis of variance (ANOVA) was calculated using the mixed procedure of SAS 9.2 (SAS Institute, 2009), taking into account the fact that strip intercropping experiments lack randomization. Thus, different spatial models were applied for the correlation of subplots within the same plot. The underlying theory, the different spatial models, as well as the procedure were described in Knörzer et al. (2010b).

In addition, the influence of different microclimates was compared within intercropped and monocropped systems in the year 2009. From May until harvest of the respective crop, wind speed, soil temperature and solar radiation values were summed. The results for intercropped species were expressed as a proportion of the corresponding monocropped results (Table 3) in order to gain a general overview of differences in total amount of microclimatic resources during co-growth. For wheat, the soil temperature sum during co-growth was calculated as (Friis et al., 1987):

$$T_{sum}(T_b) = \sum_{d=p}^{q}(T_d - T_b) \quad \vert \text{ for } T_d > T_b \quad (1),$$

where $T_{sum}(T_b)$ = soil temperature sum, T_d = soil temperature (°C), T_b = base temperature (°C) and for the period day number d = p to day number d = q.

In addition, for wheat, three sequential harvests and the final harvest were taken for the monocropping and the intercropping system. Subsequently, grain yield was correlated with each of the explanatory variables indicating that the response was quadratic rather than linear with respect to the three regressor variables (Fig. 2).

For the model approach, the DSSAT-CSM software shell version 4.5 (DSSATv4.5 Company) and therein the CERES model was applied which was developed for sole cropping systems only (Jones et al., 2003). It is a process-oriented crop model that considers soil–plant–atmosphere and management systems. It was designed to help researchers adapt and test the cropping system model itself as well as for management applications (Jones et al., 2003). The dataset collected in this case study was used to test a model extension for relay intercropping of wheat and maize. The modeling considered crop management practices such as fertilization, soil and genotype characteristics and weather data (©<Hohenheimer Klimadaten>). Phenology and growth data from both intercropped and monocropped species from both growing seasons (2007–2009) were used to estimate the cultivar coefficients (Table 1).

Whereas the genotype, soil characteristics and management did not differ between monocropped and intercropped systems in the two experimental years, the microclimate between the subplots changed.

The extension of the model's capability to simulate intercropping systems involved the inclusion of a simple shading algorithm as well as input of micro-climate modified variables such as incoming solar radiation, soil temperature, wind speed and CO_2. The coupling points for the testing of CSM-CERES models were as follows:

(1) Solar radiation was used to evaluate a shading algorithm, the output of which was then used to modify the weather input;

(2) Soil temperature measurements, N_{min} samples and nitrogen budget calculations were considered as inputs for the initial conditions to account for different N availabilities between monocropping and intercropping;

(3) CO_2 modifications were based on wind speed measurements at canopy level.

Table 1: Cultivar coefficients of the wheat variety 'Cubus' and the maize variety 'Companero' as used in the model testing

winter wheat		
parameters	description	Value
P1V	sensitivity to vernalisation	30
P1D	sensitivity to photoperiod	60
P5	grain filling duration	700
G1	kernel number per unit weight at anthesis	23
G2	kernel weight under optimum conditions	36
G3	standard stem and spike dry weight at maturity	1.0
PHINT	phylochron interval	130

maize		
parameters	description	value
P1	growing degree days from emergence to end of juvenile phase	110.0
P2	photoperiod sensitivity	0.500
P5	cumulative growing degree days from silking to maturity	700.0
G2	potential kernel number	550.0
G3	potential kernel growth rate	7.70
PHINT	phylochron interval	26.00

Solar radiation: development of shading algorithm

Among the wheat genotype parameters, CSM-CERES assumes a standard plant height of 100 cm and does not calculate specific plant height of different cultivars. For maize, plant height or plant height coefficients are absent. In addition, CSM-CERES simulates one crop at a time making it difficult to introduce an interactive or modified solar radiation algorithm for intercropping situations. As intercropping systems consist of two or more species, neighbouring plant effects cannot be ignored. Hence, an algorithm was developed to combine the neighbouring plant height and its daily shading capacity on the target species. A shading pattern based on weekly sunlight measurements was determined with regard to the height of neighbouring plants (Knörzer et al., 2010a). Incoming solar radiation of monocropped species was set to 0% shading meaning that the monocropping system was driven by the original weather data measured at the local weather station. Then the light differences of the intercrops in proportion to the monocrops were calculated. A regression analysis was performed, using the calculated shading values in percent (%) on a given day and relating them to the corresponding measured neighbouring plant height. The shading algorithm for wheat and maize resulted in the following equations:

Shading pattern within wheat:

$S_{wheat} = 0.2278 H_{maize} - 33.417$,   ($R^2 = 0.7$)   (2a)

with S_{wheat} = shading of wheat border rows in % and H_{maize} = plant height (cm) of maize.

Shading pattern within maize:

$S_{maize} = -0.003 H^3_{wheat} + 0.3909 H^2_{wheat} - 12.125 H_{wheat} - 5.0502$,   ($R^2 = 0.7$)   (2b)

with S_{maize} = shading of maize border rows in % and H_{wheat} = plant height (cm) of wheat.

The shading algorithm was only applied temporarily, during co-growth of both species taking measured data as vertex. Therefore, the weather file for both species was modified between sowing time of maize and wheat harvest. Before and after those points in time, the weather input for both intercropping and monocropping was identical. Thus, the shading curves were not extrapolated. As a result, the shading algorithm for wheat was negative, when maize was sown and no shading occurred in wheat meaning that intercropped wheat received more sunlight compared to monocropped wheat. Intercropped wheat received more sunlight in comparison to monocropped due to less intraspecific competition for light, as border rows of wheat lack competing neighbours.

Conversely, when maize grew taller than wheat, shading in intercropped wheat occurred and the output values of the shading algorithm switched to positive.

Logistic models for daily plant height as a function of days after sowing (DAS) for both species were fitted separately for the two experimental years by non-linear least squares using the NLIN procedure of the SAS system:

$$\text{Plant height} = \frac{\alpha_{ij}}{1 + \text{EXP}(\beta_{ij} + \gamma_{ij} \times \text{DAS})} \tag{3}$$

where α_{ij}, β_{ij} and γ_{ij} = parameters of species i during the year j.

The plant height was estimated from Eq. (3) and substituted for in Eqs. (2a) and (2b) to determine the shading effects. The daily solar radiation (SRAD) expressed in MJ m^{-2} d^{-1} was then calculated according to (2a) and (2b) and modified for the intercropping weather input (SRAD$_{\text{inter}}$) as follows:

$$\begin{aligned} \text{SRAD}_{\text{inter/wheat}} &= \text{SRAD}_{\text{mono}} \\ &- \left(\left(0.2278 \times \frac{\alpha_{ij}}{1 + \text{EXP}(\beta_{ij} + \gamma_{ij} \times \text{DAS})} \right) - 33.417 \right) \\ &\times \frac{\text{SRAD}_{\text{mono}}}{100} \end{aligned} \tag{4a}$$

$$\begin{aligned} \text{SRAD}_{\text{inter/maize}} &= \text{SRAD}_{\text{mono}} - \left(\left(-0.003 \times \frac{\alpha_{ij}}{1 + \text{EXP}(\beta_{ij} + \gamma_{ij} \times \text{DAS})} \right) \right)^3 \\ &+ 0.3909 \left(\frac{\alpha_{ij}}{1 + \text{EXP}(\beta_{ij} + \gamma_{ij} \times \text{DAS})} \right)^2 - 12.125 \\ &\times \left(\frac{\alpha_{ij}}{1 + \text{EXP}(\beta_{ij} + \gamma_{ij} \times \text{DAS})} \right) - 5.0502) \times \frac{\text{SRAD}_{\text{mono}}}{100} \end{aligned} \tag{4b}$$

The CSM-CERES calculates daily solar radiation according to Spitters et al. (1986) in order to model canopy photosynthesis by separating the diffuse and direct components of global radiation. Solar radiation as input variable has some influence on individual plant growth modules as well as for the soil module interface. To convert solar radiation into photosynthetically active radiation (PAR), the factor 0.5 is used (Spitters et al., 1986). Thus, an evaluated shading of e.g. 10% means that the PAR for the intercropping was reduced by 10%. The shading algorithm does not take diffuse and direct sunlight into account as such. The fact that the algorithm was used to directly modify the weather input file bypassed the problem of direct and diffuse radiation as the DSSAT-CSM models use absolute daily radiation input to calculate radiation components according to Spitters et al. (1986). The shading algorithm indirectly influences the modeled daily PAR or photon flux density as well as relative cloudiness. Measurements of solar radiation directly above the growing canopy taken to evaluate the algorithm were within the range of total radiation according to the solar radiation measurements of the local weather station.

Soil temperature

Increased top soil temperature combined with increased mineralization rate and competitive ability might lead to different N mineralization rates (Köller and Linke, 2001) of the border in comparison to the centered wheat subplots. Higher N mineralization rates were taken into account in the modeling approach by modifying initial conditions. For the N budget, soil temperature, N_{min} content and N uptake were considered, and the internal efficiency of N use (IE_i) (Eq. (5)) as well as the relative N yield total (RNT) (Eq. (6)) were calculated according to Zhang et al. (2008):

$$IE_i = \frac{Y_i}{N_i},\quad (5)$$

where Y_i = yield of crop *i*, and N_i = N uptake by crop *i*.

$$RNT = \frac{N_{wi}}{N_{ws}} + \frac{N_{mi}}{N_{ms}},\quad (6)$$

where N_{wi}, N_{ws}, N_{mi}, N_{ms} = N uptake of inter/sole wheat/maize.

A value of RNT exceeding the land equivalent ratio (LER) suggests that crops in intercropping systems were not N-efficient (Zhang et al., 2008).

Wind speed

Both a shorter and a taller species have an influence on their neighbouring species. Neighbour effects were studied in wind shelter belts placed among trees or bushes in order to decrease transpiration in arid and windy regions. Nevertheless, increased or decreased wind speed in border rows (intercropping) compared to centered rows (monocropping) might change the transpiration rate as well as the amount of CO_2 assimilation. An increased CO_2 assimilation together with increased solar radiation leads to an increased rate of photosynthesis up to the saturation point. In our experiment a slightly increased wind speed may not have increased transpiration rate as reported by Grace (1988), but may have lead to an increased CO_2 assimilation rate (Chapman et al., 1954; Wilson and Wadsworth, 1958). On a calm day, without water stress and with a moderate saturation vapor pressure deficit of the atmosphere, the plants exposed to increased wind speed may benefit from the increased CO_2 assimilation rate. According to Wilson and Wadsworth (1958), the rate of CO_2 uptake increased an average of 1.7 times when wind speed was doubled, up to a wind speed limit of approximately 1.7 m s^{-1}, after which no further increase in CO_2 uptake was produced. Wilson and Wadsworth (1958) reassessed the experiments conducted by Deneke (1931) with six different species (*Avena sativa*, *Cereus tetragonus*, *Nerium oleander*, *Polytrichum*, *Thuja gigantea*, *Tradescantia pendula*). As measured average wind speed at canopy level did not exceed 1.6 except on one day in our experiment, between May and June additional daily CO_2 supply was calculated according to Eq. (5) and added to the model using the environmental modification tool of CSM-CERES:

$$CO_{2\,additional} = \left(\frac{wind_{inter}}{wind_{mono}} \times CO_2 \times \frac{1.7}{2}\right) - CO_2 \qquad (7)$$

Introducing wind speed effects into the model by implementing an increased CO_2 amount via the environmental modification tool is crucial. Grace (1988) clearly pointed out that effects on the leaf-atmosphere boundary layer based upon wind speed are bewilderingly erratic, varying between years, locations and species. Nevertheless, CO_2 fluxes alter with wind speed within a range of a few percent. As the diffusion path for CO_2 depends on aerodynamic resistance, stomata resistance and mesophyll resistance, the immediate effects of wind speed might be rather small (Grace, 1988). Using Eq. (7) for the modeling should be considered as a first attempt to take measured microclimate changes into account and to emphasize the importance of considering more than one competition factor when modeling intercropping.

In summary, the model was tested for the monocropping and intercropping system in the years 2007–2009. The wheat and maize monocropping systems used the standard weather file with the measured daily maximum and minimum air temperature, wind speed, rainfall and solar radiation from the local weather station and an initial N input determined by N_{min} samples taken at the beginning of the growing season. For all scenarios described in the following, the wheat and maize intercropping model used the modified solar radiation (Eqs. (4a) and (4b)) weather file without directly modifying other climate factors such as air temperature or rainfall. In addition, the intercropping wheat and maize models were run with the same management inputs as the monocropping model.

In contrast to the monocropped wheat, the intercropped wheat received in scenario I a higher amount of initial N ha^{-1} as initial condition, which was calculated according to an N-balance: the N_{min} content at the beginning of the growing season plus the N uptake from intercropped wheat minus the N uptake from monocropped wheat. In both years the intercrop accumulated around 120 kg N ha^{-1} more than its monocropping equivalent.

In scenario II, the intercropped wheat received an additional amount of CO_2 instead of the additional N as initial input. The additional daily CO_2 was calculated according to Eq. (7) and the results were used as input into the 'environmental modification' tool within CSM-CERES.

In scenario III, scenarios I and II were combined. Intercropped wheat received additional N as well as additional CO_2. Only data from 2009 was used for scenarios II and III. Thus, increased soil temperature and consequently increased N availability as well as increased wind speed for intercropped wheat were taken into account. An algorithm for additional N uptake and CO_2 accumulation was not evaluated, but the CSM-CERES tools were applied. The main reason for this is because the study was to test whether those effects are important at all and, if so, to which extent. The coefficient of determination (R^2), the root mean square error (RMSE) and the model error in % were used to estimate the discrepancy between simulated and observed values and the model fit.

RESULTS

Crop yield and microclimate

Yield, yield components and N uptake of wheat and maize for the years 2007/08 and 2008/09 are shown in Table 2. The results from the two experimental years showed that intercropping effects are restricted to a few border rows or to field boundaries. Wheat benefited from its neighbouring maize plants due to its advanced stage of development and its competitive ability in acquiring nutrients and light. Average grain yield of intercropped wheat was around 3 t ha^{-1} higher than monocropped wheat due to a higher number of tillers, which was on average more than one quarter higher. Dry matter of intercropped wheat was about 7 t ha^{-1} higher than monocropped wheat.

Maize showed no differences in grain yield and dry matter accumulation between border and centered rows but tended to develop more ears per plant in border rows. Mean grain yield of both years and systems was approximately 10 t ha^{-1} and mean dry matter yield was 26 t ha^{-1}.

Table 2: Yield and yield components of intercropped and monocropped wheat and maize grown in Germany in the growing season 2008/09. Statistical analysis was done by fitting a linear model with fixed position effects. Means in the same column sharing no letter are significantly different at α = 0.05 according to a t-test

Cropping system	Tiller m^{-2} (no.)	Grain yield (kg ha^{-1})	Dry matter yield (kg ha^{-1})	N uptake (kg ha^{-1})
Wheat				
Intercropping	706^b	11975^b	24424^b	333^b
2-4 m	554^a	9110^a	17246^a	216^a
Monocropping	562^a	9070^a	17114^a	212^a
Cropping system	**Ears plant^{-1} (no.)**	**Grain yield (kg ha^{-1})**	**Dry matter yield (kg ha^{-1})**	**N uptake (kg ha^{-1})**
Maize				
Intercropping	1.4^b	10984^a	26262^a	301^a
3/4 row	1.2^a	10215^a	24645^a	296^a
5/6 row	1.3^a	10552^a	27127^a	327^b
Monocropping	1.3^a	10549^a	26166^a	324^b

Accumulated solar radiation calculated according to the shading algorithm indicated that intercropped wheat accumulated about 15% more light than monocropped wheat (Table 3). Nevertheless, the light distribution over time differed. Intercropped wheat received more light than its monocropping equivalent until mid-June and a similar amount of light until the end of June, which corresponded to flowering and grain filling (Fig. 1). In contrast, intercropped maize received approximately 15% more light than monocropped maize, despite the fact that intercropped maize was shaded from the taller neighbouring wheat plants at the beginning of the maize growing season. The early reduction in incoming solar radiation was compensated for after wheat harvest, when intercropped maize was sunlit. Thus, the total amount of solar radiation may have been less important than the light distribution throughout the growing season.

Table 3: Comparison between the influence of different microclimate within intercropped and monocropped systems in the year 2009 from May until harvest of the respective crop. The amount of wind speed, soil temperature and solar radiation was summed. The results for monocropped species was stated as 100% and accordingly the proportion of the corresponding intercropped species calculated

	Grain yield (%)	Soil temperature (%)	Wind speed (%)	Solar radiation (%)
wheat$_{inter}$	132	110	107	115
maize$_{inter}$	104	105	97	115

The correlation of microclimate parameters with grain yield confirmed that the influence of top soil temperature, wind speed as well as solar radiation was not linear over the growing season. Instead there was a quadratic response with a maximum (Fig. 2). The influence of those parameters increased during grain filling and decreased during ripening. There was a strong correlation between solar radiation and soil temperature ($R^2 = 0.999$).

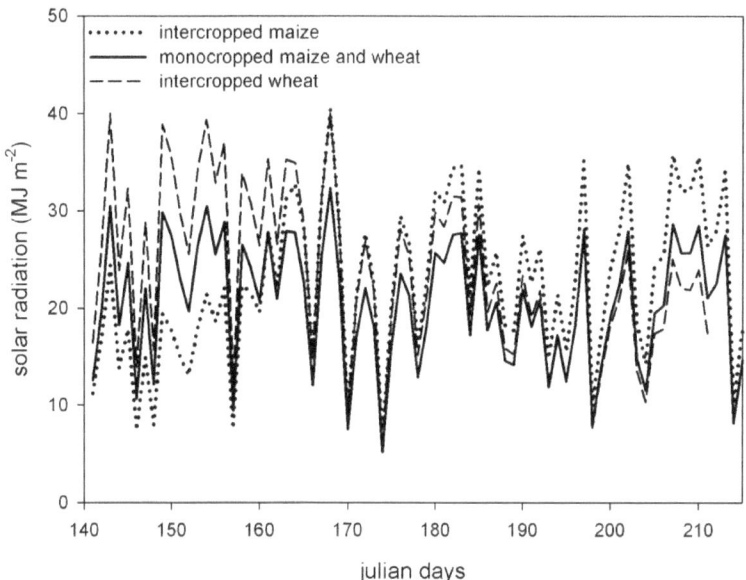

Figure 1: Incoming solar radiation (MJ m^{-2}) of monocropped and intercropped maize and wheat during May, June and July in the year 2009. Solar radiation of monocropping was taken from local weather station measurements. Solar radiation of intercropping was calculated according to the evaluated shading algorithm

For maize, soil temperature was similar between border and centered rows during the whole measuring period. The sum of soil temperature between May and harvest date for intercropped maize was 5% higher than monocropped maize (Table 3).

In contrast, soil temperature decreased from wheat border to wheat centered rows by 1–2°C on average. In comparison to the monocrop, the intercrop showed a soil temperature increase of 10%, which was equivalent to a discrepancy of soil temperature degree sum of 123. Soil temperature degree sum of monocropped wheat was 1197 and 1320 for intercropped wheat between May and harvest date.

Plant N uptake until ripe dough stage increased in wheat border rows and decreased in maize border rows. In both years, wheat accumulated an additional 120 kg N ha^{-1} in border rows compared to centered rows. Increased soil temperature may increase N mineralization and N availability (Köller and Linke, 2001). Measured N_{min} plotted against soil temperature showed an inverse correlation with soil temperature being steeper for monocropped ($R^2 = 0.79$) than for intercropped ($R^2 = 0.87$) wheat (Fig. 3).

Table 4: Land equivalent ratio (LER), relative nitrogen yield total (RNT) and internal efficiency of nitrogen use (IE) of monocropped and intercropped maize and wheat in the year 2009

Intercropping system	LER[a]	RNT[b]	IE[c]
Maize/wheat	1.36	1.39	Maize$_{mono}$ = 32.5 Maize$_{inter}$ = 36.5 Wheat$_{mono}$ = 37.2 Wheat$_{inter}$ = 33.6

[a]land equivalent ratio (Wubs et al., 2005)
[b]relative N yield total (Zhang et al., 2008)
[c]internal efficiency of N use (Zhang et al., 2008)

For the same amount of available N in the soil, a higher soil temperature was required in the intercropping system. Moreover, when the N_{min} content in the monocropped and intercropped wheat was the same, the intercropped wheat accumulated more N than monocropped wheat. Until ripe dough stage, monocropped wheat accumulated 212 kg N ha^{-1}, while intercropped wheat accumulated 333 kg N ha^{-1}. Mineralization increased in the intercropping system and the increased N amount in the soil became available and taken up by the intercropped wheat.

Nevertheless, the intercropping system did not increase the N efficiency in comparison to the monocropping system (Table 4). The RNT value for the wheat/maize intercropping system was 1.39 and the LER was 1.36. The internal efficiency of N use for intercropped wheat was 33.6 and for monocropped wheat it was 37.2 indicating that monocropped wheat had a slightly improved N-use efficiency compared to intercropped wheat. In contrast, the internal N-use efficiency for intercropped maize in this system was 36.5 and for monocropped maize it was 32.5.

In maize, wind speed did not differ between monocropping and intercropping irrespective of the component crop (Fig. 1). Wind speed ranged between 0.0 and 2.0 m s^{-1} with an average of 0.5 m s^{-1}. The wind speed of the intercropped maize was 97% of that in the sole maize crop (Table 3) between May and harvest date.

For wheat, the taller component crop maize had a greater influence on the border rows of the cropping systems. Taller neighbouring plants influence wind speed at canopy level of shorter target plants. Even if wind speed is defined by a much broader range of parameters, there appeared to be a correlation (Fig. 4) between wind speed at canopy level and neighbouring-plant height. At the beginning of the growing season, wind speed increased in border rows until maize had the same height as wheat. When maize grew taller than the component crop, wind speed was reduced within border rows of wheat compared to the monocropping situation. The average wind speed in wheat between May and July was 0.4 m s^{-1} for both intercropping and monocropping, and intercropped wheat experienced a 7% higher wind speed than the monocrop.

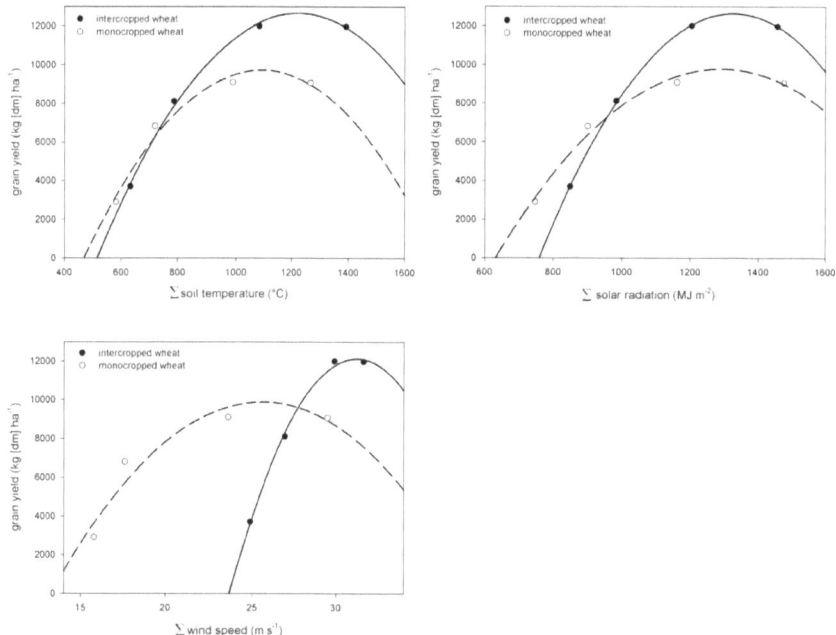

Figure 2: Influence of different microclimate parameters and different cropping systems on wheat grain yield during the growing season

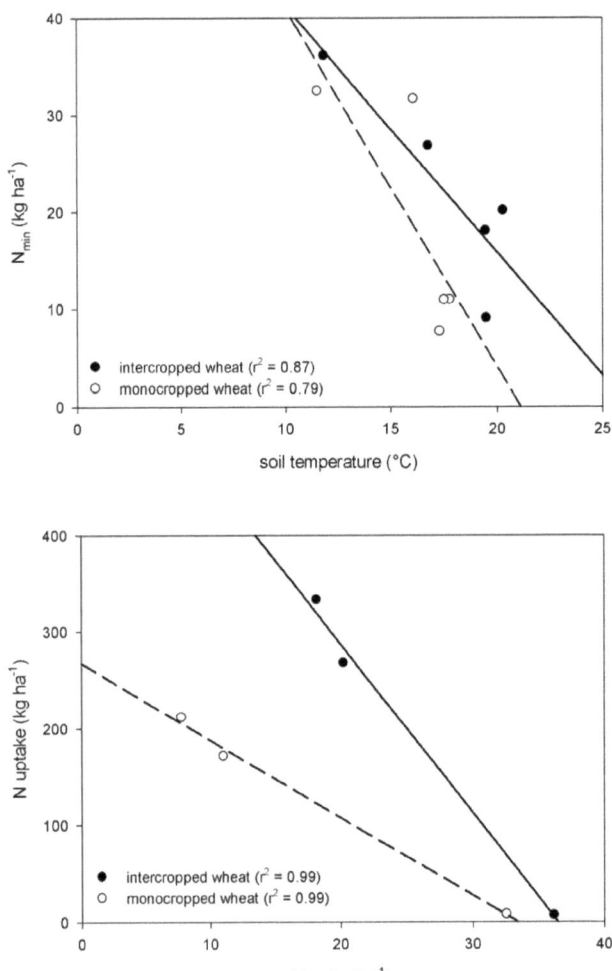

Figure 3: N_{min} content vs. soil temperature (top) and N uptake vs. N_{min} (below) in monocropped and intercropped wheat during the growing season 2009

Simulation and model testing

Monocropped and intercropped wheat

Each intercropping model scenario used a solar radiation input that was modified according to the evaluated and respective shading algorithm during co-growth. The result of these modified weather inputs is illustrated in more detail in Fig. 1.

The modeling results of scenario I (modified solar radiation and initial N content) for intercropped and monocropped wheat in the years 2007–2009 showed a good fit between simulated and observed grain yield as well as dry matter accumulation (Fig. 5, Table 5), although the model did not simulate the slope of the sigmoid curve for grain yield adequately. Average observed grain yield for the years 2007–2009 was 10.6 t ha^{-1}, with intercropping yielding 12.2 t ha^{-1} and monocropping yielding 9.1 t ha^{-1} on average. Average simulated grain yield was 10.7 t ha^{-1} with intercropping reaching 12.5 t ha^{-1} and monocropping 8.9 t ha^{-1}. The RMSE for grain yield at maturity was 300 kg ha^{-1} with a model error of 2.8%. Average observed as well as simulated dry matter yield was 20.7 t ha^{-1}, with a RMSE of 779 kg ha^{-1} and a model error of 3.8%. Intercropped wheat had 24.4 t ha^{-1} observed and 23.7 t ha^{-1} simulated dry matter yield. Monocropped wheat had 17.0 t ha^{-1} observed and 17.6 t ha^{-1} simulated dry matter yield. As a result, the R^2 value was 0.98 for grain yield as well as for dry matter yield.

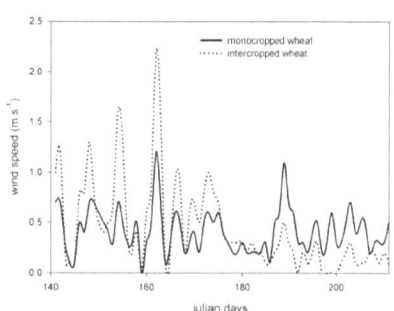
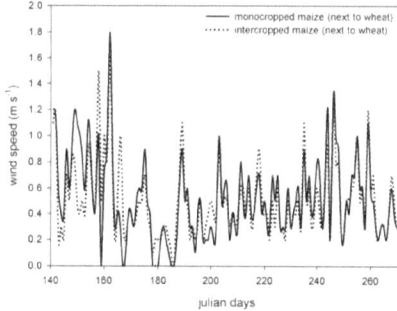

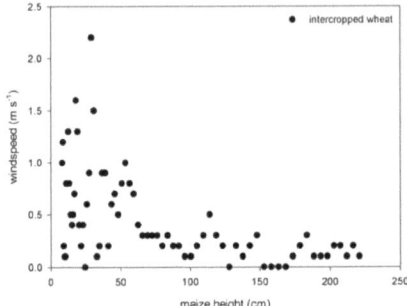

Figure 4: Measured wind speed (m s^{-1}) above the canopies of intercropped and monocropped maize and wheat in the year 2009 between May and harvest date of respective crops. Wind speed at intercropped wheat canopy level in dependence on neighbouring maize height

In addition, for the year 2009, different scenarios were tested: the baseline model with the modified solar radiation and the initial N input for simulating intercropping; the model with the modified solar radiation and the environmental modifications concerning CO_2; and finally the combination of modified solar radiation, modified initial N input and changed CO_2 concentrations (Table 5). Scenario I as well as scenario II showed a good fit between observed and simulated grain and dry matter yield. The RMSE for scenario I was 423 kg ha^{-1} (model error = 4.0%) for grain and 656 kg ha^{-1} (model error = 3.2%) for dry matter yield. Replacing the N input with additional CO_2 in scenario II led to a similar model fit. RMSE for grain yield was 524 kg ha^{-1} (model error = 5.0%), RMSE for dry matter yield was 511 kg ha^{-1} (model error = 2.5%). CSM-CERES overestimated grain and dry matter yield when applying both additional N and CO_2, as was done within scenario III. Model error increased to 14.4% for dry matter accumulation and 16.9% for grain yield. Using the environmental modifications as stand-alones together with modified solar radiation, the model simulated the increase in yield of intercropped wheat adequately, but some effects may be overestimated and need to be recalibrated.

Table 5: Model testing results for monocropped and intercropped wheat 2007–2009 when using different scenarios. The first scenario used modified solar radiation (SRAD) and a higher amount of initial N ha^{-1} (N). The second scenario replaced additional N input with increased CO_2 (CO_2) uptake, and in a final scenario, intercropped wheat received additional N as well as additional CO_2

	Mean (obs.) (kg ha^{-1})	Mean (sim.) (kg ha^{-1})	RMSE (kg ha^{-1})	Model error (%)
Scenario I (SRAD+N) 2007-2009				
Intercropping:				
Dry matter:	24464	23746	748	3.1
Grain yield:	12186	12454	398	3.3
Monocropping:				
Dry matter:	16984	17550	745	4.4
Grain yield:	9062	8912	160	1.8
Combined intercropping/monocropping:				
Dry matter:	20724	20670	779	3.8
Grain yield:	10624	10700	300	2.8
Scenario I (SRAD+N) 2009				
Combined intercropping/monocropping:				
Dry matter:	20768	20348	656	3.2
Grain yield:	10522	10700	423	4.0
Scenario II (SRAD+CO$_2$) 2009				
Combined intercropping/monocropping:				
Dry matter:	20768	21168	511	2.5
Grain yield:	10522	10774	524	5.0
Scenario III (SRAD+N+CO$_2$) 2009				
Combined intercropping/monocropping:				
Dry matter:	20768	22930	3000	14.4
Grain yield:	10522	11674	1781	16.9

Monocropped and intercropped maize

Maize intercropped with wheat suffered at the beginning of the growing season due to the competitiveness of wheat. But because wheat was harvested approximately three months earlier than maize, maize showed a recovery-compensation growth after harvest of wheat, as already described by Li et al. (2001). As a result, maize grain yields as well as dry matter accumulation were similar between monocropping and intercropping. The monocropped model was run taking the measured N_{min} values into account. In contrast, the intercropped model was run with N stress at the beginning of the growing season, starting with less N content in the soil than the monocropped equivalent in order to account for the increased competitiveness of wheat for N acquisition in comparison to maize. Due to the recovery-compensation growth, the simulation of both monocropped and intercropped maize should be similar to each other, although the intercropping model was run with the modified solar radiation file. The observed steep slope of grain filling was not simulated adequately. Nevertheless, the simulation of grain and dry matter yield at maturity showed a sufficient fit between measured and observed yields (Fig. 6).

The model error for grain yield was 9.6% and for dry matter accumulation 8.7% with a RMSE for grain yield of 945 kg ha^{-1} and for dry matter yield of 2168 kg ha^{-1} (Table 6). Mean measured grain yield and dry matter yield was 9.8 t ha^{-1} and 25 t ha^{-1}, respectively. The simulation results were 10 t ha^{-1} for mean grain yield and 23.5 t ha^{-1} for mean dry matter accumulation.

Table 6: Model testing results for monocropped and intercropped maize 2007–2009. Scenario I used modified solar radiation (SRAD) and a higher amount of initial N ha^{-1} (N)

	Mean (obs.) (kg ha^{-1})	Mean (sim.) (kg ha^{-1})	RMSE (kg ha^{-1})	Model error (%)
Scenario I (SRAD+N) 2007-2009				
Intercropping:				
Dry matter:	24779	23842	1684	6.8
Grain yield:	9918	10481	940	9.5
Monocropping:				
Dry matter:	25310	23066	2562	10.1
Grain yield:	9726	9523	950	9.8
Combined intercropping/monocropping:				
Dry matter:	25045	23454	2168	8.7
Grain yield:	9822	10002	945	9.6

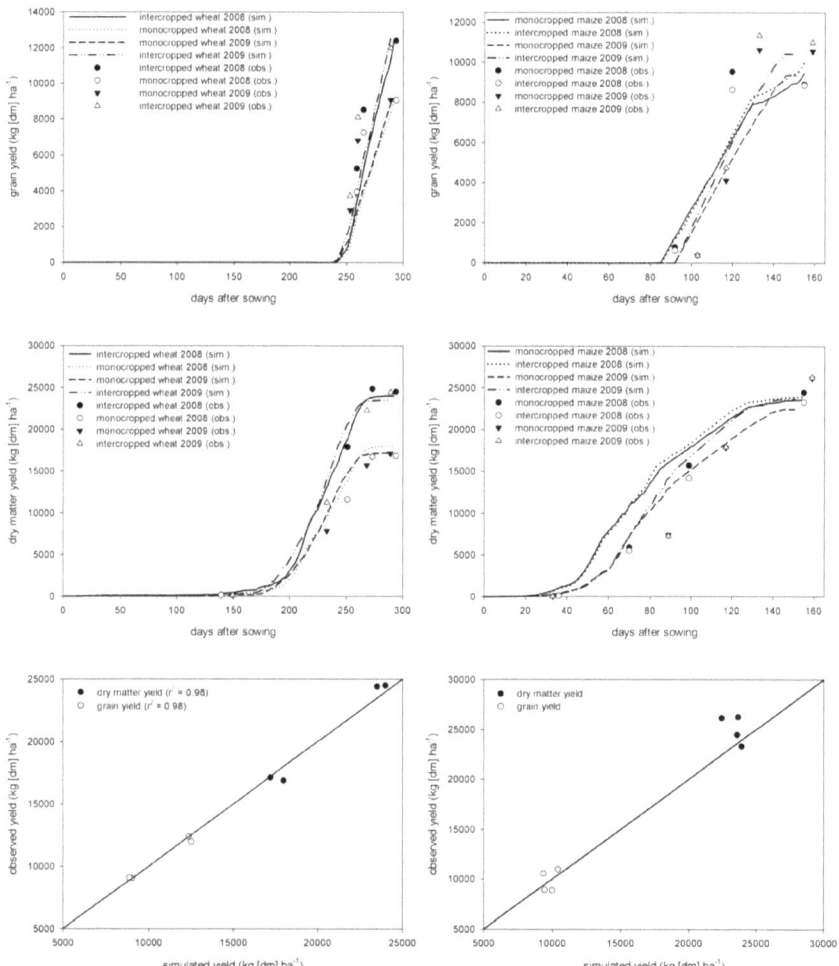

Figure 5: Simulated and observed grain yield (top) and dry matter accumulation (middle) for intercropped and monocropped wheat 2007–2009 in dependence on days after sowing. Observed vs. simulated yield (below) indicate the model fit

DISCUSSION

Competition for solar radiation is the key component in intercropping systems if growth is not limited by water or nutrient supply. Most studies dealing with modeling of intercropping or interspecific competition have directed their attention to shading effects or radiation algorithms (Baumann et al., 2002; Brisson et al., 2004; Sellami and Sifaoui, 1999; Tsubo and Walker, 2002; Wiles and Wilkerson, 1991). The turbid layer medium analogy (Brisson et al., 2004; Sinoquet et al., 2000) as well as the approach to reduce leaf area index (LAI) of the understorey species while the dominant species reaches a particular height and shading occurs (O'Callaghan et al., 1994; Rossiter and Riha, 1999) were two common attempts to deal with the different amounts of incoming sunlight in intercropping systems. Both approaches need to consider plant height of the neighbouring species and LAI to allow an inversion of dominance in height between species during the crop cycle (Corre-Hellou et al., 2009). Dominance or the dominant components in intercropping systems are defined as species with initial size advantage (Andersen et al., 2007). For interspecific competition, plant height is the trait most commonly associated with competitiveness (Zerner et al., 2008).

There were two main reasons for evaluating a shading algorithm and not adopting common Beer's law approaches for modeling interspecific competition in order to simulate relay intercropping within the DSSAT-CSM framework. Firstly, the CSM-CERES model has no plant coefficients for maize height and assumes only a standard wheat height, hence height effects cannot directly be introduced into CSM-CERES. Nevertheless, the model has often proven to be robust in the simulation of monocropping scenarios. Jones et al. (2003) gave a brief overview of various types of applications of the DSSAT-CSM crop models and example references that describe these applications in detail, organized by continent on which the studies were conducted. In addition, the CERES and CROPGRO models integrated in DSSAT are generic models containing several species and making it easier to integrate new species. Species variation is needed to model the great variety of existing intercropping systems. Therefore, integrating into CSM-CERES a simple shading algorithm that takes neighbouring plant height into account seems to be a promising approach for extending the model from monocropping to intercropping.

Secondly, the turbid layer medium analogy, which is a common approach for intercropping and mixed plant growth models, bears little resemblance to a relay intercropping system. In these systems, where the competition effects occur in a few rows, and where the development of crops is staggered, plant growth cannot be modeled and simulated according to Beer's law. Within point-based models, the mass balance is kept constant, meaning there is a defined partitioning pool of light, nutrients and water. Hence, beneficial effects such as increased solar radiation or N availability violating this principle of constant mass flows cannot be modeled, meaning that there is no winner or looser within the system. Naturally, the more resources the dominant species catches, the less will be caught by the understorey species. In effect, the wheat–maize relay intercropping showed that wheat border rows benefited from intercropping with maize, and maize did not suffer.

In principle, detected differences within the wheat plot as well as the maize plots were restricted to the first few rows reflecting the intercropping situation and giving evidence of intercropping being a field border effect. Thus, modeling intercropping could be defined as modeling border effects. Simulation was carried out for the monocropping situation within the centered subplots and the subplots with the shortest distance from the border.

Applying a simple shading algorithm for the intercropping situation in order to estimate the proportion of shading in comparison with the monocropping situation and as a function of neighbouring plant height, allowed additional as well as reduced solar radiation to be taken into account for the wheat border rows during the growing season. In addition, the shading algorithm altered the distribution of incoming solar radiation during the growing season, but not necessarily the total amount.

As a result, grain and dry matter yield of wheat and maize were simulated in good accordance with the measured grain and dry matter yield. Model error of grain and dry matter yield for both species was under 10%. There was a tendency for the model to simulate grain yield adequately but to slightly underestimate dry matter yield. For intercropped wheat, a linear shading algorithm seemed to be appropriate (Knörzer et al., 2010a); for maize, a polynomial shading algorithm was tested. A submodel was not needed. Instead the solar radiation (SRAD) in the standard weather file was modified in daily steps according to changing neighbouring plant height and percentage shading over the growing season without extrapolating the curves. The advantage of that method is that it could be adapted for all species combinations, and a simpler approach could be an adequate surrogate. For plant height, sigmoid curves are easy to evaluate and normally they show a good fit. In addition, the data needed for the shading algorithm can be evaluated in the field and the later estimation of the shading is based upon measured data. It is also independent of crop coefficient and LAI. LAI measurements are difficult to realize in intercropping or border rows as most of the measurement tools require a monocropping situation to measure LAI adequately (e.g. LI-COR Plant Canopy Analyzer), and destructive methods require more effort and a larger plot size due to the increased number of plants cut for LAI measurements during the growing season. Nonetheless, plant growth correlates better with light interception than with incoming light and LAI is an important determinant of light capture. Thus, LAI will not be disregarded in the simulation, but in the algorithm. CSM-CERES calculates LAI in relation to photosynthesis, but LAI is not an explicit input variable for simulation.

Variations in microclimate, e.g. soil temperature and wind speed, within an intercrop as well as their influence over time also merit attention. In intercropping, resource distribution in space and time seems to be more important than the total amount of resources (Tsubo et al., 2001). With optimized resource distribution the overall efficiency can be equivalent to monocropping. According to the estimates, intercropped wheat and maize received approximately 15% more solar radiation than the monocropped equivalent. Intercropped wheat assimilated additional sunlight up to mid-June and less sunlight as from the end of June. As a result, intercropped wheat received more sunlight during the entire vegetative phase and half of the grain filling phase. Intercropped maize assimilated less solar radiation at the beginning of the growing season for around two months. Before flowering and ear setting, intercropped maize obtained additional sunlight. Evans and Wardlaw (1976) stated that shading and reduced assimilate production will have the least effect on yield if they occur during vegetative growth, and conversely, the greatest effect during the reproductive growth phase. In their review of the competition effects of solar radiation in intercropping, Keating and Carberry (1993) noted that if a crop is harvested before the major yield determining period of the companion crop, yield of the subsequent crop may be little affected. Our regression analysis of cumulated microclimate influence at four sequential time harvests confirmed that the influence is not linear, as Tsubo et al. (2001) suggested for cumulative PAR, but quadratic (Fig. 2). The recovery compensation growth for maize intercropped with wheat, described by Li et al. (2001), as well as the N uptake recovery of cotton intercropped with wheat, described by Zhang et al. (2008), might be explained by the light distribution over time. Then, intercropping should be light timed and timing of shade is one of the most important agronomic factors in intercropping (Tsubo and Walker, 2004).

Our results confirmed that intercropped maize does not suffer from being intercropped with wheat, yielding as high as monocropped maize over two cropping seasons. In addition, wheat benefited from being intercropped with maize, and grain yield increased by up to 32%. Gao et al. (2009) reported a yield increase in wheat/maize relay intercropping from 40% to 70%, and total grain yield rose by 39–98% as compared to maize and wheat yields in monocultures. In contrast, Wang et al. (2009) argued in their review of wheat cropping systems in China that farmers hesitate to extend the relay intercropping of wheat and maize because yield is reduced by 5–10%.

Some studies showed that intercrops use soil-nutrients more efficiently than sole crops (Inal et al., 2007; Li et al., 2004; Zhang and Li, 2003; Zuo et al., 2004), however agronomic N-use efficiency may also be decreased by intercropping (Zhang et al., 2008). Intercropped wheat accumulated 120 kg more N ha^{-1} than its monocropping equivalent and intercropped maize accumulated less kg N ha^{-1}. The internal N-use efficiency was more favourable for monocropped wheat and intercropped maize. The LER was notably increased by about 36% without increasing the RNT. The wheat/maize system was N-efficient and as a result N dynamics have to be taken into account when simulating intercropping scenarios.

In a first modeling approach which used the linear shading algorithm for wheat, Knörzer et al. (2010a) showed that modifying only the incoming solar radiation did not fully explain the yield increase in intercropped wheat grain and dry matter. Only when the initial conditions were changed, taking the higher amount of N cycling in the system into account, could the yield increase be explained. Not only was the competitiveness of wheat in comparison to maize an important reason for the higher N accumulation but nitrogen availability should also be taken into consideration. Due to the increased top soil temperature (5 cm) within the first rows – 2°C difference on average – the mineralization of N might be favored and intercropped wheat might receive more N than its monocropped equivalent. The previous crop, sugar beet, had an average N surplus of about 55 kg N ha^{-1} (Reisch and Knecht, 1995) after harvest. The management system was a reduced tillage system. Lower soil temperature and a reduced mineralization at the beginning of the growing season in comparison to tillage systems are typical for reduced tillage systems (Köller and Linke, 2001). Hence, the increased soil temperature within the field boundary might lead to a better N supply of wheat in the first rows. Nutrient supply is much more dependent on microclimate, water content and soil properties (Raynaud and Leadley, 2005) and cannot be restricted to competition only.

Some microclimate effects are small, subtle or non-existent, while spatial and climate variability and the heterogeneity of plant populations can be considerable (Meinke et al., 2002). Quantifying those effects may be difficult but should not be neglected. To date, the influence of wind speed on crop yield is more often studied in windbreak or shelter zones than in intercropping. Song and Wei (1998) found that a reduction of 1 m s^{-1} of wind speed caused by shelter increased crop yield by 19 g m^{-2}. Evapotranspiration is sensitive to wind speed and modified wind speed and turbulence alter the exchanges of heat, water vapor and CO_2 (Meinke et al., 2002). Increased or decreased wind speed in border rows compared to centered rows might change the transpiration rate as well as the CO_2 assimilation rate. Wheat border rows were exposed to a higher wind speed until mid-June. After June, wheat border rows were exposed to reduced wind speed. CO_2 assimilation rate may be increased during wet and cooler months, whereas transpiration was reduced during the hot summer months. Both effects were difficult to quantify, but may led to an improved crop performance. The model testing showed that modified CO_2 as well as modified N inputs improved the model fit for intercropping systems. Both N and CO_2 had a similar effect on crop yield and appeared to be exchangeable.

The underlying mechanisms of different wind speed regimes in monocrops and intercrops are complex and difficult to quantify. Surface resistances, evaporation, convection, and turbulence intensities play a significant role in wind shelter, as demonstrated in detail by Grace (1988). They have to be considered on single leaf-levels rather than being generalized for a dense plant stand. In conclusion, the effects as stand-alones may be overestimated in this study and need to be analyzed more closely. Nevertheless, those effects have to be taken into account when simulating interspecific competition. This study does not intend to validate an intercropping approach for CSM-CERES but is a first step towards integrating a solar competition algorithm and using specific CSM-CERES tools in order to account for microclimate effects. Furthermore, the evaluated shading algorithm is a first approach to avoid using common competition submodels such as modified Beer's law approach which are restricted to mixed or at least row intercropping (Knörzer, in press). Strip or relay intercropping modeling needs to be revised in terms of better taking into account the beneficial effects which can result in the violation of resource pools.

Integrating a simple shading algorithm into CSM-CERES seemed to be promising as wheat and maize grain and dry matter yield could be simulated adequately in both the intercropping and the monocropping systems. However, more than one competition effect, e.g. solar radiation needs to be factored in. In further modeling studies, microclimate influences such as soil temperature and wind speed should be considered.

ACKNOWLEDGEMENTS

The authors' research topic is embedded in the International Research Training Group of the University of Hohenheim and China Agricultural University, entitled "Modeling Material Flows and Production Systems for Sustainable Resource Use in the North China Plain". We thank the German Research Foundation (DFG) and the Ministry of Education (MOE) of the People's Republic of China for their financial support. We also would like to thank Mrs. Nicole Gaudet for her diligent proof reading of the manuscript.

REFERENCES

Andersen, M.K., Hauggaard-Nielsen, H., Weiner, J., Jensen, E.S., 2007. Competitive dynamics in two- and three-component intercrops. Journal of Applied Ecology 44, 545–551.

Baumann, D.T., Bastiaans, L., Goudriaan, J., Van Laar, H.H., Kropff, M.J., 2002. Analysing crop yield and plant quality in an intercropping system using an eco-physiological model for interplant competition. Agricultural Systems 73, 173–203.

Brisson, N., Bussiére, F., Ozier-Lafontaine, H., Tournebize, R., Sinoquet, H., 2004. Adaptation of the crop model STICS to intercropping, theoretical basis and parameterisation. Agronomie 24, 409–421.

Caldwell, R.M., 1995. Simulation models for intercropping systems. In: Sinoquet, H., Cruz, P. (Eds.), Ecophysiology of Tropical Intercropping. INRA editions, Paris, pp. 353–368.

Chapman, H.W., Gleason, L.S., Loomis, W.E., 1954. The carbon dioxide content of field air. Plant Physiology 29, 500–503.

Corre-Hellou, G., Faure, M., Launay, M., Brisson, N., Crozat, Y., 2009. Adaption of STICS intercrop model to simulate crop growth and N accumulation in peabarley intercrops. Field Crops Research 113, 72–81.

Deneke, H., 1931. Über den Einfluß bewegter Luft auf die Kohlensäureassimilation. Jahrbuch für wissenschaftliche Botanik 74, 1–32.

Evans, L.T., Wardlaw, I.F., 1976. Aspects of the comparative physiology of grain yield in cereals. Advances in Agronomy 28, 301–359.

Friis, E., Jensen, J., Mikkelsen, S.A., 1987. Predicting the date of harvest of vining peas by means of air and soil temperature sums and node counting. Field Crops Research 16, 33–42.

Gao, Y., Duan, A., Sun, J., Li, F., Liu, Z., Liu, H., Liu, Z., 2009. Crop coefficient and water-use efficiency of winter wheat/spring maize strip intercropping. Field Crops Research 111, 65–73.

Grace, J., 1988. Plant response to wind. Agriculture, Ecosystems and Environment 22/23, 71–88.

Harris, D., Natarajan, M., 1987. Physiological basis for yield advantage in a sorghum/groundnut intercrop exposed to drought, 2. Plant temperature, water status and components of yield. Field Crops Research 17, 273–288.

Hoogenboom, G., Wilkens, P.W., Tsuji, G.Y. (Eds.), 1999. DSSAT v 3, volume 4. University of Hawaii, Honolulu, Hawaii.

Inal, A., Gunes, A., Zhang, F., Cakmak, I., 2007. Peanut/maize intercropping induced changes in rhizosphere and nutrient concentrations in shoots. Plant Physiology and Biochemistry 45, 350–356.

IUSS Working Group WRB, 2007. World Reference Base for Soil Resources 2006, first update 2007. World Soil Resources Reports No. 103. FAO, Rome.

Jones, J.W., Hoogenboom, G., Porter, C.H., Boote, K.J., Batchelor, W.D., Hunt, L.A., Wilkens, P.W., Singh, U., Gijsman, A.T., Ritchie, J.T., 2003. The DSSAT cropping system model. European Journal of Agronomy 18, 235–265.

Keating, B.A., Carberry, P.S., 1993. Resource capture and use in intercropping: solar radiation. Field Crops Research 34, 273–301.

Knörzer, H. Designing, modeling, and evaluation of improved cropping strategies and multi-level interactions in intercropping systems in the North China Plain. PhD thesis, Univesität Hohenheim.

Knörzer, H., Graeff-Hönninger, S., Guo, B., Wang, P., Claupein, W., 2009. The rediscovery of intercropping in China: a traditional cropping system for future Chinese agriculture – a review. In: Lichtfouse, E. (Ed.), Springer Series: Sustainable Agriculture Reviews 2: Climate Change, Intercropping, Pest Control and Beneficial Microorganisms,. Science + business media, Berlin, pp. 13–44.

Knörzer, H., Graeff-Hönninger, S., Müller, B.U., Piepho, H.-P., Claupein, W., 2010a. A modeling approach to simulate effects of intercropping and interspecific competition in arable crops. International Journal of Information Systems and Social Change 1 (4), 44–65.

Knörzer, H., Müller, B.U., Guo, B., Graeff-Hönninger, S., Piepho, H.-P., Wang, P., Claupein, W., 2010b. Extension and evaluation of intercropping field trials using spatial models. Agronomy Journal 102, 1023–1031.

Köller, K.H., Linke, C., 2001. Erfolgreicher Ackerbau ohne Pflug: Wissenschaftliche Ergebnisse – Praktische Erfahrungen. 2. neu überarb. u. erw. Aufl., Frankfurt a.M. DLG, Germany.

Li, S.M., Li, L., Zhang, F.S., Tang, C., 2004. Acid phosphatase role in chickpea/maize intercropping. Annals of Botany 94, 297–303.

Li, L., Sun, J., Zhang, F., Li, X., Yang, S., Rengel, Z., 2001. Wheat/maize or wheat/soybean strip intercropping. I. Yield advantage and interspecific interactions on nutrients. Field Crops Research 71, 123–137.

Meinke, H., Carberry, P.S., Cleugh, H.A., Poulton, P.L., Hargreaves, J.N.G., 2002. Modelling crop growth and yield under the environmental changes induced by windbreaks.1. Model development and validation. Australian Journal of Experimental Agriculture 42, 875–885.

Meng, E.C.H., Hu, R., Shi, X., Zhang, S., 2006. Maize in China; Production Systems, Constrains, and Research Priorities. International Maize and Wheat Improvement Center (CIMMYT), Mexico.

O'Callaghan, J.R., Maende, C., Wyseure, G.L.C., 1994. Modelling the intercropping of maize and beans in Kenya. Computers and Electronics in Agriculture 11, 351–365.

Raynaud, X., Leadley, P.W., 2005. Symmetry of belowground competition in a spatially explicit model of nutrient competition. Ecological Modeling 189, 447–453.

Reisch, E., Knecht, G., 1995. Betriebslehre. Stuttgart. Ulmer, Germany.

Rossiter, D.G., Riha, S.J., 1999. Modeling plant competition with the GAPS object-oriented dynamic simulation model. Agronomy Journal 91, 773–783.

SAS Institute, 2009. The SAS System for Windows Release 9. 2. SAS Institute, Cary, NC.

Sellami, M.H., Sifaoui, M.S., 1999. Modelling solar radiative transfer inside the oasis; experimental validation. Journal of Quantitative Spectroscopy & Radiative Transfer 63, 85–96.

Sinoquet, H., Rakocevic, M., Varlet-Grancher, C., 2000. Comparison of models for daily light partitioning in multispecies canopies. Agricultural and Forest Meteorology 101, 251–263.

Song, Z., Wei, L., 1998. The Correlation between Windbreak Influenced Climate and Crop Yield. International Research Centre, Ottawa, DRC: Library: Documents: Agroforestry Systems in China.

Spitters, C.J.T., Tousaint, H.A.J.M., Goudriaan, J., 1986. Separating the diffuse and direct component of global radiation and its implications for modeling canopy photosynthesis, Part I: components of incoming radiation. Agricultural and Forest Meteorology 38, 217–229.

Tsubo, M.,Walker, S., 2002. AModel of radiation interception and use by maize-bean intercrop canopy. Agricultural and Forest Meteorology 110, 203–215.

Tsubo, M.,Walker, S., 2004. Shade effects on Phaseolus vulgaris L. intercropped with Zea mays L. under well-watered conditions. Journal of Agronomy and Crop Science 190, 168–176.

Tsubo, M., Walker, S., Mukhala, E., 2001. Comparison of radiation use efficiency of mono-/intercropping systems with different row orientations. Field Crops Research 71, 17–29.

Wang, F., He, Z., Sayre, K., Li, S., Si, Y., Feng, B., Kong, L., 2009. Wheat cropping systems and technologies in China. Field Crops Research 111, 181–188.

Wiles, L.J., Wilkerson, G.G., 1991. Modeling competition for light between soybean and broadleaf weeds. Agricultural Systems 35, 37–51.

Wilson, J.W., Wadsworth, R.M., 1958. The effect of wind speed on assimilation rate – a reassessment. Annals of Botany 22, 285–290.

Wubs, A.M., Bastiaans, L., Bindraban, P.S., 2005. Input levels and intercropping productivity: exploration by simulation. Plant Research International, note 369, Wageningen.

Zerner, M.C., Gill, G.S., Vandeleur, R.K., 2008. Effect of height on the competitive ability of wheat with oats. Agronomy Journal 100, 1729–1734.

Zhang, F., Li, L., 2003. Using competitive and facilitative interactions in intercropping systems enhances crop productivity and nutrient-use efficiency. Plant and Soil 248, 305–312.

Zhang, L., Spiertz, J.H.J., Zhang, S., Li, B., Van der Werf, W., 2008. Nitrogen economy in relay intercropping systems of wheat and cotton. Plant and Soil 303, 55–68.

Zuo, Y., Liu, Y., Zhang, F., Christie, P., 2004. A Study on the improvement of iron nutrition of peanut intercropping with maize on nitrogen fixation at early stages of growth of peanut on a calcareous soil. Soil Science and Plant Nutrition 50, 1071–1078.

9 Chapter VI:

Evaluation and performance of the APSIM crop growth model for German winter wheat, maize and fieldpea varieties within monocropping and intercropping systems

PUBLICATION VI:

Knörzer, H., Lawes, R., Robertson, M., Graeff-Hönninger, S., and Claupein, W. (2011): Evaluation and performance of the APSIM crop growth model for German winter wheat, maize and fieldpea varieties in monocropping and intercropping systems. Journal of Agricultural Science and Technology 5 (12), pre-print version.

The basic idea behind chapter VI was to use two different models concerning their intercropping simulation approach to test both approaches with the same dataset and to compare the outcome. The intention was less to determine an inferior or superior intercropping modeling approach, but to improve the knowledge how to model interspecific competition within strip intercropping and to study the contribution and extent of different competition algorithms on crop performance. Otherwise, without such a comparison or suggestion, the question is reliable, why evaluating another competition model instead of improving already existing ones (chapter II)? For that study or comparison, the APSIM crop growth model was chosen because of three reasons: 1) APSIM is based upon DSSAT in its development and previous version and thus, in both models monocropped species are modeled in a similar manner and with similar assumptions; 2) APSIM has been used in several multiple cropping studies instead of just one case study so far and seemed to be a useful and validated tool; 3) The APSIM competition algorithm and approach is based upon Beer's law, often used in other models studies (chapter II). As APSIM has been used predominantly for Australian crop growing conditions, the model had to be calibrated for temperate European zones conditions in a first step. Results from the previous chapter were taken then to compare the differences between the DSSAT and the APSIM intercropping modeling approach.

9. Chapter VI

Evaluation and performance of the APSIM crop growth model for German winter wheat, maize and fieldpea varieties in monocropping and intercropping systems

H. Knörzer[1], R. Lawes[2], M. Robertson[2], S. Graeff-Hönninger[3] and W. Claupein[3]

[1]Institute of Biobased Products and Energy Crops (340b), University of Hohenheim, 70593 Stuttgart, Germany
[2]Centre for Environment and Life Sciences, CSIRO Perth, Floreat WA 6014, Australia
[3]Institute of Crop Science (340a), University of Hohenheim, 70593 Stuttgart, Germany

Article from the Journal of Agricultural Science and Technology 5 (12): pre-print version, with permission, copyright © 2011, David Publishing Company.

ABSTRACT

Competition for solar radiation between plants grown in multi-species cropping systems can severely limit crop production of individual species within that system. There are various approaches for modeling light interception within mixed-cropping and row or strip intercropping systems. To extend the knowledge about model behavior and different model approaches under interspecific competition conditions, the APSIM model was evaluated and calibrated for field experiments previously described and simulated by DSSAT. Initially the APSIM plant model was successfully modified to simulate wheat, maize and fieldpea monocultures in the European agro-ecological zone. Once calibrated, the APSIM model was then used to simulate a strip relay intercropping maize/wheat and maize/fieldpea system. In DSSAT, a shading algorithm was introduced to modify the daily weather input in order to take competition for solar radiation into account. In addition microclimate influences were modified for the relay intercropping system. In contrast, APSIM simulates interspecific competition using the canopy module, which is based on a modified Beer's law for multi-component or mixed canopy conditions, and a routine that defines soil water and nutrient supply for each species component in a daily alternating manner. After a re-evaluation of the model regarding a minimum change of crop coefficients and variables, APSIM was able to simulate dry matter and grain yield of German maize, winter wheat and fieldpea varieties adequately. However, APSIM is a point-based model, and many of the processes that influence strip cropping cannot be accommodated by adjusting Beer's Law alone. So far none of the tested frameworks successfully modeled strip or relay intercropping. The processes governing growth in the numerous and very diversifying intercropping systems are complex and at this point in time have not been captured in sufficient detail. Further modifications to DSSAT and APSIM are required to successfully simulate strip or relay intercropping. We discuss some of the key changes that need to be made to these models to simulate strip intercropping systems.

Keywords: APSIM, competition, DSSAT, intercropping, modeling, solar radiation

INTRODUCTION

Adjusting cropping systems in order to increase their efficiency is a global issue. High yield and sustainability are the catchphrases of production in the 21st century, and agricultural production has to solve the balancing act between ecology and economy. Therefore, the demands on farmers, consultants and researchers are rising and production modes are changing. Nevertheless, solutions need to be found, accepted and implemented locally in order to be successful. The use of modeling and simulation tools supports the acceleration of research attainments and the understanding of cropping systems. The application of crop growth models has increasingly become a scientific tool for the analysis of cropping systems. Those models have mostly been evaluated for monocrops. With the paradigm of sustainability in mind, the modeling of mixed cropping and intercropping systems is of increasing interest. In the intercropping research context the following benefits appear repeatedly [14]: maximized land use, several harvests per year, yield stability, increased resource use efficiency, reduced soil erosion and leaching, reduced pests and diseases. The simulation of both systems, intercropping and monocropping, requires robust and carefully evaluated and validated models. There are various approaches to modeling intercropping and interspecific competition [5][15]. Most of these approaches deal with the modeling of competition for solar radiation in mixed species canopies [12]. In this paper, two issues of modeling are addressed.

1.) A brief methodology of calibrating a crop growth model for monocropping and intercropping scenarios is shown. For this, the APSIM model (Agricultural Production Systems Simulator) [17][18] was chosen.

2.) A comparison between two different modeling intercropping approaches was carried out to gain further insight into the methodology models used to capture the competitive processes that govern plant growth and development in these systems. This is combined with a critical review of persisting shortcomings concerning the modeling of strip and relay intercropping systems. DSSAT (Decision Support System for Agrotechnology Transfer) [11] was chosen for comparison with APSIM.

The process-oriented APSIM Vs 7.1 (www.apsim.info/Wiki/APSIM-Documentation.ashx) crop growth model was used to model monocropping and intercropping experiments. APSIM is a dynamic soil-plant-atmosphere model, similar to the DSSAT Vs 4.5 crop growth model, which allows modeling and simulation of crop and pasture production, residue decomposition, soil water and nutrient flow and management influences.

Previous versions of the cereal models – e.g. APSIM-Nwheat [2] -, included into APSIM were mainly based upon the DSSAT CERES models. In addition, the evaporation algorithm developed by Ritchie [25] is used in both models. Thus, DSSAT and APSIM share a common set of modeling features. Nevertheless, APSIM was revised [1][8][23] and the version 7.1, used in this study, differed from earlier versions. For example, changes were made to the number of plant stages in the wheat module.

In contrast to DSSAT, APSIM includes a competition model that allows the simultaneous simulation of two crops competing for solar radiation, soil water and nutrients to a limited extend. Additionally, APSIM and DSSAT differ in modeling incoming solar radiation and photosynthetic active radiation (PAR). DSSAT uses the approach described by Spitters et al. [30]; APSIM uses the Beer's law approach described by Monsi and Saeki, [21]. In addition, the simulation of relay intercropping with the DSSAT model [16] pursues a different approach than with the APSIM model [6]. In DSSAT, a shading algorithm as well as modified microclimate influences were taken into account in this study by modifying the daily weather input with particular regard to solar radiation and the initial conditions within the relay intercropping system. In contrast, APSIM simulates interspecific competition using the canopy module based on a modified Beer's law for multi-component or mixed canopy conditions and a routine that defines soil water and nutrient supply for each species component in a daily alternating manner. Various canopy layers are defined starting at the top of the tallest canopy which is equal to the plant height of the dominant species in the system. The fraction of light transmitted out of the top layer can be calculated, and this in turn is the fraction entering the next layer below [4].

In this paper, we demonstrate a methodology for calibrating the APSIM model for a monocropping system in temperate European zones, a region where it had not previously been used. We then compare and contrast the two different modeling intercropping approaches employed in APSIM and DSSAT [16] to gain further insight into the methodology the models use to capture the competitive processes that govern plant growth and development in these systems. Finally we critically review the shortcomings concerning the modeling of intercropping systems.

MATERIAL AND METHODS
Field experiments 2007 to 2009

Experiments were conducted in southwest Germany at the University of Hohenheim experimental station 'Ihinger Hof' during the years 2007 to 2009. The station is located 48.46°N and 8.56°E and has an average temperature of 7.9°C per year, an average rainfall of 690 mm per year and 1674 average sunshine hours per year. Dominant soils are silty loamy medium-deep para-brown colored soils over loess, classified as *Orthic Luvisols* [10].

The experiment comprised a maize/wheat relay intercropping system as well as a maize/fieldpea intercropping system, each within a complete block design with four replications. Thus, each replication contained complete blocks of both systems and each system's replication consisted of four plots (Figure 1). Two plots were used for sequential harvests and weekly measurements during the growing season and another two plots were used for the final harvest. The two species were planted in an alternate pattern. Each plot was 10 x 10 m² for wheat and for pea, and 12 x 10 m² for maize, and included five subplots (5 x 2 x 10 m²) for wheat and pea, and eight subplots (8 x 1.5 x 10 m²) for maize. Within those subplots, data was collected to detect crop performance differences between different distances from the plot border. The plots were large enough for the central subplot to effectively represent a monocropping system. Row orientation was from north to south.

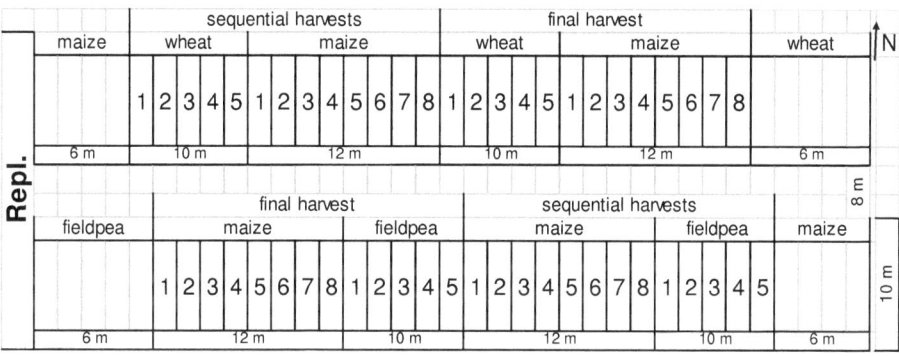

Figure 1: Experimental design of the strip/relay intercropping systems maize/winter wheat and maize/fieldpea during the years 2007 to 2009. Field trials were conducted in southwest Germany

In both years, the previous crop was sugar beet and soil preparation and sowing was done by a reduced tillage system. Plant protection was carried out according to 'Good Agricultural Practice'. The wheat variety 'Cubus' was sown in October 2007 and 2008 with a row spacing of 13 cm and a plant density of 300 plants per m². Maize ('Companero') was sown in May 2008 and 2009 with a row spacing of 75 cm and a plant density of 10 plants per m². The fieldpea variety 'Hardy' was sown between the end of March and the beginning of May in the years 2008 and 2009 with a row spacing of 13 cm and a plant density of 70 plants per m². Harvest time of wheat and pea was at the end of July. Maize was harvested in October.

Wheat was fertilized with 160 kg N ha^{-1}, split into three applications (60/60/40) of Nitro-chalk. Maize was fertilized once with 160 kg N ha^{-1} (ENTEC). Fieldpea, being a leguminous plant, was not fertilized.

Three sequential harvests were carried out as square meter cuts, and grain yield, dry matter and nitrogen (N) concentration of plants were analyzed for the years 2007 - 2009. Nitrogen concentration was determined with the NIRSystems 5000 (ISI-Software, USA). Growing stages according to the German BBCH [19] scale and plant height were reported on a weekly basis. After the final harvest, yield and yield components such as thousand kernel weight (TKW), tiller number, ears or pods per plant as well as N concentration and N uptake were determined for all crops.

Model set-up

For the monocropping model approach, the APSIM crop growth model version 7.1 was applied. Datasets for soil characteristics and daily weather (©<Hohenheimer Klimadaten>) for the years 2007 to 2009 were identical to those used for the DSSAT modeling approach [16]. Initial nitrogen content for the soil model was set according to mineralized nitrogen content (N_{min}) measurements taken for each crop at the beginning of the growing season. Management input variables such as sowing date, sowing density, sowing depth, fertilizer application as well as fertilizer amount were set according to field experiment management and documentation. Each crop and each year was simulated with separate model runs. Thus, the initial soil variables varied from one year to the next.

For the intercropping simulation, the evaluated plant variety specific coefficients within the monocropping system were used. The companion crop sown later was added to the component sown earlier and linked through the canopy module [4] without changing soil, weather or management input.

Evaluation of the winter wheat variety model

Within the APSIM model, differences in performance of various wheat varieties are based on differences in overall accumulated thermal time units for maturity, photoperiodic sensitivity and vernalization requirements taking thermal time units for the individual and different growing stage durations as a fixed setup. So far, the APSIM model has mainly been used for Australian and African temperate, semi-arid, tropical and subtropical simulation studies and has proved to be a robust model for simulating crop growth with particular regard to water shortage and drought [1][9][13][27]. Nevertheless, Asseng et al. [2] showed that the APSIM Nwheat model, a previous version of the APSIM model used in this study, can also be used to model crops grown in a European environment. Thus the variables in the APSIM Vs 7.1 wheat module were modified to simulate German high-yielding wheat varieties. This was achieved by modifying the three variables (thermal time unit requirements, photoperiodic and vernalization sensibility) used to simulate different wheat varieties within the model. No fundamental processes were modified.

The main problem within the modeling of German wheat varieties in comparison to Australian varieties was that dry matter was overestimated for the German varieties and, by contrast, grain yield was underestimated. The performance of total biomass could be adequately simulated, but the allocations to stem and leaf dry matter on the one hand and grain yield on the other hand seemed to be inappropriate. In addition, the simulated duration of floral initiation was longer than that observed and the beginning of flowering and grain filling was later than that observed. As a result, to improve the model fit, further evaluations of crop coefficients and variables had to be performed with special regard to biomass partitioning, dry matter allocation and the duration and beginning of individual crop growth stages.

The model was evaluated stepwise, using data from German field experiments from the years 2007 to 2009 and the wheat variety 'Cubus' grown under monocropping conditions [16]. The variety was evaluated using the data from the years 2007 – 2008, and the modified coefficients and variables were tested with the data from 2008 – 2009. Changes within the wheat model were partly made by using coefficients similar to those used in the DSSAT crop growth model for the wheat variety 'Cubus' and are shown in Table 1.

Table 1: Results of the stepwise model evaluation of APSIM, coefficient and variable modifications for the German wheat variety 'Cubus'

Phenology	Coefficient / variable	Default			Calibrated		
Maturity type	maturity type	580			610		
Thermal time calculation	x_temp_units	0	26	34	0	16	35
Phylochron interval	y_node_app_rate	95			130		
Biomass partitioning	Biomass Partitioning	6	6.9	7	4	4	3
Dry matter allocation	Fraction of remaining dm allocated to leaves	0.6	0.6	0.42	0.5	0.5	0.52
	Fraction of dm allocated to pod	0.33	0.33	0.33	0.1	0.3	0.3
Senescence	x_dm_sen_frac_leaf	1			2		
	y_dm_sen_frac_leaf	1			0.5		
	node_sen_rate units	60			58		
	fr_lf_sen_rate	0.035			0.033		

The first step was to test the model using the given variables for the individual varieties, especially maturity type. Thermal time units for maturity were changed from 580 to 610, thus elongating the growing season and increasing the overall dry matter yield. The next step was testing the model performance after changing the base temperature for grain filling in the thermal time calculation from 26°C to 16°C [7]. This is accommodate the fact that European winter wheat varieties are much more adapted to cool temperatures not only during the winter months, but also in spring and early summer. Accordingly to the calculated thermal time units from field data collected in 2008 (120) and the DSSAT cultivar coefficients (130), values concerning phylochron intervals were changed from the standard 95 to 130 [16]. Biomass partitioning coefficients as well as coefficients for the dry matter allocation to leaves and pods were modified to reduce stem and leaf dry matter and to increase grain yield. Only those partitioning coefficients were changed which determine allocations during the flowering and grain filling period to postpone the allocation and prolong the life cycle of the crop. This change allows the crop to develop more photosynthetically active tissue, increases the total amount of carbohydrates and finally increases grain yield. Observed data from separated stem, leaf and ear dry matter during the growing season were taken as basic assumptions. In a last step, senescence coefficients were evaluated to ensure that the shift of dry matter allocation was eventually used for increased grain yield instead of increased leaf dry matter. Earlier senescence of leaves can lead to a reallocation of assimilates from leaves to grains and hence increase grain yield. After each evaluation step, the model was run and the following variables were compared to detect the differences in model behavior and the impact of the modifications made on wheat phenology: dry matter (biomass), actual above-ground dry matter (dlt_dm), potential change in live plant LAI (dlt_lai_pot), potential change in LAI allowing for stress (dlt_lai_stressed), change in number of leaves (dlt_leaf_no), potential leaf number (dlt_leaf_no_pot), extractable soil water in different soil layers (esw_layr(1), esw_layr(2), esw_layr(3)), plant water uptake (ep), N in grain (grain_n), grain number (grain_no), weight of grain (grain_wt), size of each grain (grain_size), LAI (lai), N demand of plant (n_demand), N uptake (n_uptake), demand for NO_3 (no3_demand), NO_3 available to plants (no3_tot), root depth (root_depth), growing stage (stage), soil water supply (sw_supply), soil water demand (sw_demand), soil water deficit in different layers (sw_deficit(1), sw_deficit(2)) and grain yield (yield). Starting point was an appropriate Australian wheat variety ('tennantstart') for comparison, adding step by step the modified maturity type ('maturity'), thermal time calculation ('thermtime'), phylochron interval ('PHINT'), biomass partitioning ('biomasspart'), dry matter allocation to leaves ('dmleaves'), dry matter allocation to pods ('dmpod') and lastly senescence ('cubusend').

Evaluation of the maize variety model

To introduce a new variety into the APSIM maize model, several variables had to be defined or calibrated. Most of those variables define the accumulation of thermal time units within the individual growing stages, plant development and the switching from the vegetative to generative phase. Maize varieties differ in these traits, especially when adapted to and bred for different purposes and climatic zones. Thus, thermal time units evaluated for Australian and African maize varieties within the APSIM model have to be modified in order to model maize varieties bred and grown in temperate zones. Main differences are increased water supply, decreased incoming solar radiation, adaptation to lower average temperatures and adaptation to the duration of the growing season of maize grown in temperate zones in comparison to maize grown in semi-arid, tropical or subtropical zones. Again, as in the wheat model, the main challenge for the simulation of maize growth in temperate zones is to model the possibility of high yielding varieties in those areas with yields exceeding those in major Australian crop production regions. Simulation runs with default maize varieties within the APSIM model showed that these were not able to cope with German soil, weather and input variables and, as a result, crop growing stages were not simulated adequately. In particular flowering and grain filling had to occur earlier for the German variety 'Companero', grown under monocropping conditions. In most cases, the given default maize varieties did not even ripen. Thus, the coefficients and variables for growth stage appearance, development and duration had to be modified.

Similar to the wheat model evaluation, the maize model evaluation was done stepwise, using data from German field experiments from the years 2008 and 2009. The variety 'Companero' was evaluated with the data from the year 2008, and the changed coefficients and variables were tested with the data from 2009. Changes within the maize model were partly made by using coefficients similar to those used in the DSSAT model for the same variety and are shown in Table 2.

Table 2: Results of the stepwise model evaluation of APSIM; coefficient and variable modifications for the German maize variety 'Companero'

Phenology	Coefficient / variable	Default	Calibrated
Phylochron interval	leaf_app_rate1	65	25
Thermal time units	tt_emerg_to_endjuv	250	300
	tt_flower_to_maturity	550	750
	tt_flag_to_flower	50	70
	tt_maturity_to_ripe	1	5
Grain number	head_grain_no_max	850	580
Grain growth	grain_gth_rate units	11	7.7

First, the phylochron interval was changed from 65 to 25 according to DSSAT (25) and to calculated thermal units in the year 2008 (38). Once the crop was established, leaf appearance rate occurred rather fast and flowering as well as grain filling started earlier allowing the crop to develop fast within a shorter growing season in temperate zones and in addition, allowing the maize plant to have an adequate duration of grain filling in order to achieve high dry matter and grain yield. The decreased time for the plant to develop in the vegetative phase led to a decreased number of kernels per ear and a decreased grain growth rate unit later on. Thus, those variables were modified from 850 to 580 (DSSAT = 550) and 11 to 7.7 (DSSAT = 7.7), respectively. In addition, thermal time units for the individual plant stages were adjusted in order to simulate growth stage durations and maize phenology adequately.

After each evaluation step, the model was run and the following variables were compared in order to detect the differences in model behavior and the impact of the modifications done on maize phenology: biomass, dlt_dm, dlt_lai_pot, dlt_lai_stressed, dlt_leaf_no, ep, grain_n, grain_no, grain_wt, grain_size, lai, n_demand, n_supply_soil, no3_demand, no3_tot, no3_uptake(1), no3_uptake(2), stage, sw_supply, sw_demand, sw_deficit(1), sw_deficit(2) and yield. Starting point was an appropriate Chinese maize variety ('zhongdan2start') for comparison, adding step by step the modified phylochron interval ('PHINT'), thermal time units ('thermtime'), potential grain number ('grainno') and grain growth units ('companeroend').

Evaluation of the fieldpea variety model

In comparison to the wheat and the maize model within APSIM, the German fieldpea variety 'Hardy' within a monocropping system was simulated appropriately using a default Australian variety. The legume model seemed to be more robust with regard to plant-soil-atmosphere interferences than the cereal model. Moreover, there is a much wider range of plant varieties than with legumes due to more extensive breeding efforts. In Germany there are fewer than ten different varieties of fieldpea. Nevertheless, an Australian variety could be used to simulate yield of a German fieldpea crop. The timing of the progression through the various plant stages was adequately simulated by the default variety, but final grain yield was overestimated. Total biomass production was also adequately simulated, but the final partitioning between biomass and grain yield needed revision as 'Hardy' accumulated more biomass in an early stage than simulated, and observed grain yield was lower than simulated (Table 3). Both experimental years reflected a typical fieldpea performance and growth for German conditions with pea growing fast up to a height of 90 cm until flowering, but breaking down rapidly afterwards to a plant height of 20 to 30 cm, mainly due to the widespread occurrence of near wilt (*Fusarium oxysporum f. sp. Pisi*) followed by leaf and stem necrosis. As APSIM is not able to simulate plant pests and diseases, dry matter yield at flowering stage was taken for the model testing instead of final dry matter. Thus, dry matter accumulation was taken as a reference value before the disease occurred.

Table 3: Results of the stepwise model evaluation of APSIM, coefficient and variable modifications for the German fieldpea variety 'Hardy'

Phenology	Coefficient/variable	Default		Calibrated	
Leaf number and area senescence	node_sen_rate units	46.6		20	
Biomass Partitioning	frac_leaf_units	0.43	0.4	0.35	0.33

After the two evaluation steps, the model was run and the following variables were compared in order to detect the differences in model behavior and the impact of the modifications made on pea phenology: biomass, yield, dlt_dm, change in retranslocation cohort 1 dry matter (cohort1retranslocationwt), dlt_lai_stressed, dlt_leaf_no, dlt_leaf_no_pot, ep, esw_layr(1), grain_wt, lai, leaf_no, number of senesced leaves per square meter (leaf_no_sen), N demand of plant (n_demanded(1), n_demanded(2)), N supply (n_supply_soil), N uptake (n_uptake), senesced dry matter (senescedwt), stage, sw_demand, sw_deficit(1) and sw_supply. Starting point was an appropriate Australian fieldpea variety ('parviestart') for comparison, adding the modified senescence ('senescence') and biomass partitioning ('hardyend') step by step.

RESULTS

Simulation of the winter wheat experiment

The stepwise model evaluation showed a steady improvement and lastly a good fit between observed and simulated wheat dry matter and grain yield (Table 4, Figure 2). In both years model derivation for those traits was below 10%.

Table 4: Results of the observed and simulated grain and dry matter yield of the winter wheat variety 'Cubus' in the years 2007 – 2009 after the modification of the APSIM wheat model

	Simulated (kg ha^{-1})	Observed (kg ha^{-1})	Δ Difference (kg ha^{-1})
Year 2007/08			
Grain yield	9067	9054	13
Dry matter harvest	17488	16854	634
Year 2008/09			
Grain yield	8677	9070	393
Dry matter harvest	17522	17114	408
Ø Years 2007 - 209			
Grain yield	8872	9062	190
Dry matter harvest	17505	16984	521

To increase dry matter and grain yield in order to reflect yield potential in temperate European zones, the modification of the maturity type in combination with the thermal time calculation was necessary. Both variables showed a similar result in increasing yield potential, especially grain yield potential, by modeling an extended growing season which was in accordance with the documented growing stages through the growing season.

The observed length of time between wheat sowing and harvest was 300 to 310 days. The default Australian variety simulated around 290 cropping days, the modification in maturity type and thermal time calculation increased this to 310 cropping days. Whereas simulated daily dry matter accumulation was not influenced by the stepwise variation of all variables and coefficients, potential change in live plant leaf area index (LAI), potential change in LAI allowing stress and potential leaf number were influenced. Maturity type as well as thermal time calculation reduced those traits compared to the default simulation. Moreover, all modifications combined had no impact on the simulation of overall LAI development. Changes in thermal time units and biomass partitioning during flowering and grain filling resulted in similar LAI development with a reduced green leaf area potential and leaf area being less susceptible to stress. With fewer but vital green leaves, the same LAI was achieved by reducing leaf biomass production and instead increasing grain yield. The reduction of biomass partitioning and dry matter allocation coefficients on behalf of grain yield can be verified by measured stem-, leaf-, ear-ratio during flowering and grain filling. The ratio from stem+leaf:ear changed from 1.6:1 to 0.3:1 within one month, and ears showed to be a strong sink. Measured harvest index was rather high at approximately 0.55. Stem and leaf production as well as remaining dry matter allocation to stem and leaf were significantly reduced in favor of ear growth and development. Additionally, modified coefficients for earlier senescence of leaves may lead to a reallocation of assimilates from leaves to grains and hence increase grain yield. Phylochron interval, biomass partitioning and senescence were the important factors for different simulation scenarios with regard to grain number, grain weight and grain protein content. There were no differences however for grain size. As dry matter allocation to leaves and stems was reduced and senescence started earlier, assimilated nitrogen and reallocated assimilates were shifted into grains, increasing not only grain yield per se, but also grain number, protein content and grain weight. With an observed 30000-31000 kernels per ha, the model still underestimated the grain number, but improved after modification simulating 22000-25000 instead of the previous 14000-17000 kernels per ha. In addition, observed protein content was between 13-14%. Without model modification, APSIM simulated 12-13%; after modification, APSIM simulated 14% grain protein content. Hence, the re-evaluation of the APSIM wheat model for simulating German wheat varieties showed that the model was able to simulate phenology effects and connected physiology effects adequately.

Finally, stress factors such as limitations in water and nitrogen supply and their impact on the default as well as the modified variety were compared. Soil water supply, soil water deficit and soil water demand within different soil layers as well as soil nitrate demand and content, for example, did not differ between the default and the stepwise modified varieties. The model modifications had no influence on those variables.

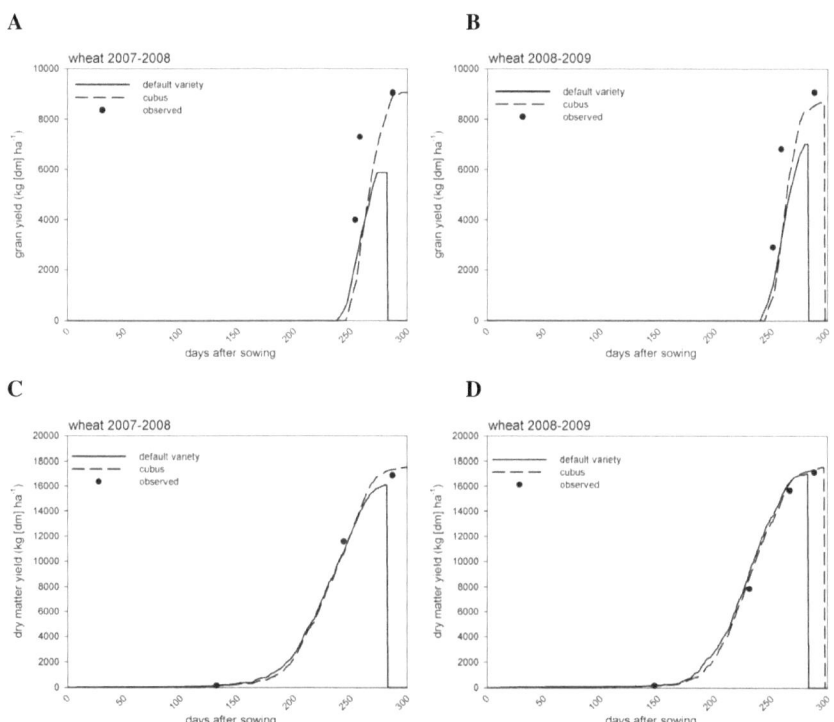

Figure 2: Measured and simulated grain (A/B) and dry matter yield (C/D) of the winter wheat variety ‚Cubus' during the growing seasons 2007/08 (a/c) and 208/09 (b/d) before (default variety) and after (cubus) the modification of the APSIM wheat model

Simulation of the maize experiment

Simulation results after the re-evaluation of the APSIM maize model showed a good fit between simulated and observed dry matter and grain yield (Table 5, Figure 3). The difference between observed and simulated average grain yield for both experimental years was 152 kg ha^{-1} and 55 kg ha^{-1} for average dry matter yield.

Table 5: Results of the observed and simulated grain and dry matter yield of the maize variety 'Companero' in the years 2008 – 2009 after the modification of the APSIM maize model

	Simulated (kg ha^{-1})	Observed (kg ha^{-1})	Δ Difference (kg ha^{-1})
Year 2008			
Grain yield	8944	8988	44
Dry matter harvest	23685	24375	690
Year 2009			
Grain yield	10216	10476	260
Dry matter harvest	27888	27089	799
Ø Years 2008/09			
Grain yield	9580	9732	152
Dry matter harvest	25787	25732	55

Modifications were needed to simulate yield height adequately. Most of the model modifications were made within the variety range and opportunities offered by APSIM. This means the determination of the crop remaining in the individual plant stages is based upon thermal time accumulation units. Once the maize plant was established, leaf tip occurrence happened rather fast as German maize varieties are adapted and bred in order to ripen within less than 200 growing days, can cope with a reduced amount of incoming solar radiation compared with tropical and subtropical regions and to achieve high yields. Thus, reduced thermal units for the phylochron interval boosted the amount of biomass and especially grain yield, but reduced the duration of the growing season, particularly the duration of the grain filling to harvest phase, drastically. As a result, grain yield was overestimated by the model after the first modification step. On the other hand, the occurrence of growing stages was modeled more adequately in comparison to the default variety except grain filling duration. Hence, a subsequent evaluation of thermal time units for flowering date and grain filling duration was necessary. With the modified thermal time units, not the duration of the whole growing season per se, but only the flowering and grain filling phase could be adjusted. Changes within the phylochron interval and the thermal time unit variables led to an increased production of daily biomass earlier in season due to an increase in daily leaf number development, daily leaf area potential and final leaf area. As a result, the vegetative phase was accelerated and the plant changed from vegetative to generative phase earlier in the season without the expense of biomass production. Nonetheless, the shortening of the vegetative phase reduced the ability of the plant to develop a high grain number potential. Modifying that coefficient led to a better model fit for the trait grain yield as it had been overestimated by the model before this evaluation step. The model still overestimated the thousand kernel weight (TKW) and assumed a higher proportion of nitrogen allocated to grain due to the increased stem and leaf biomass early in the season. The observed TKW was between 250 and 300 g. The simulated TKW ranged between 400 and 500 g. A reduction in grain growth rate simulated a final grain weight of about 350 g.

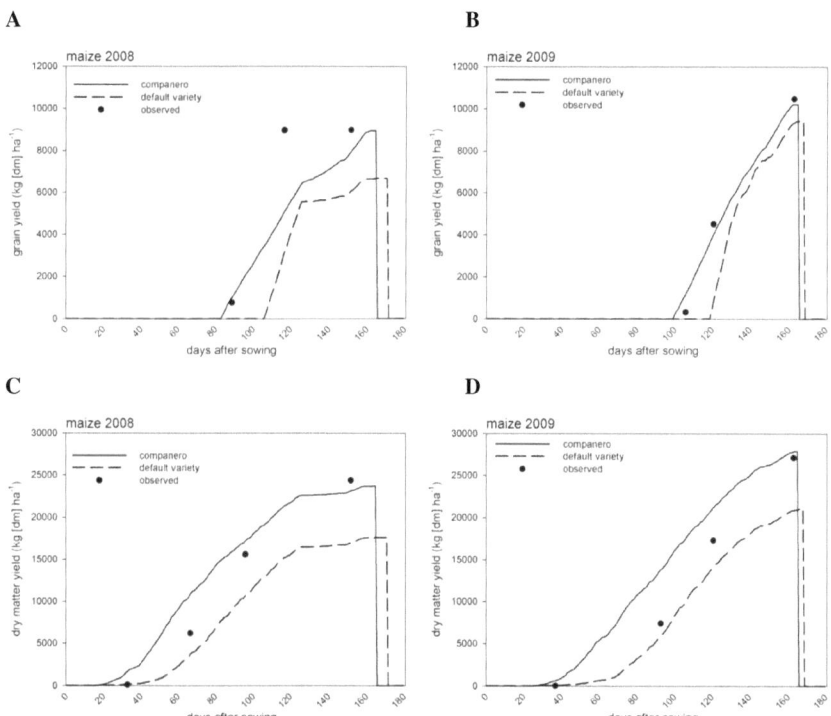

Figure 3: Measured and simulated grain (A/B) and dry matter yield (C/D) of the maize variety 'Companero' during the growing seasons 2008 (a/c) and 2009 (b/d) before (default variety) and after (companero) the modification of the APSIM maize model

In accordance with the accelerated vegetative phase of the adjusted in comparison to the default variety, the model simulated soil water and nitrogen demand earlier in the growing season with the tendency for the adjusted variety to have an increased demand for soil water and nitrogen. Water shortage is a less limiting factor for maize grown in Germany than in major Australian crop production zones.

Simulation of the pea experiment

Only a few modifications were necessary to adjust the APSIM fieldpea model to German fieldpea performance and growing conditions. Fieldpea is susceptible to waterlogging and to fungal diseases under muggy conditions, both likely to occur during the growing season in Germany. As a result, the pea leaf biomass declines rapidly after pod setting and within pod filling. Otherwise the APSIM was not able to fill the gap between dry matter accumulation and grain yield under those specific German pea growing conditions. The observed biomass accumulation was accelerated within 'Hardy' compared with the simulation, but observed grain yield was lower than simulated. After modifying biomass partitioning and senescence coefficients for the default APSIM variety, the model was able to cope with those conditions and to simulate dry matter accumulation until flowering and grain yield after harvest adequately (Table 6). The difference between simulated and observed results was under 400 kg ha^{-1} for each year and trait.

Table 6: Results of the observed and simulated grain and dry matter yield of the fieldpea variety 'Hardy' in the years 2008 – 2009 after the modification of the APSIM fieldpea model

	Simulated (kg ha^{-1})	Observed (kg ha^{-1})	Δ Difference (kg ha^{-1})
Year 2008			
Grain yield	4408	4738	330
Dry matter yield flowering	2224	2619	395
Year 2009			
Grain yield	5462	5187	275
Dry matter yield flowering	2758	3033	275
Ø Years 2008/09			
Grain yield	4935	4963	28
Dry matter yield flowering	2491	2826	335

The main priority was to set up the model in that way that until flowering sufficient biomass was simulated without simulating increased final grain yield by inducing earlier senescence. The final amount of senesced dry matter did not differ between the default and the modified variety, but the slope of the curve differed, as it rose earlier and less steep as the default. As a result, the nitrogen demand for the modified variety was reduced at the end of the growing season. The model reduced daily dry matter production as well as leaf area once the senescence coefficients had been changed, especially at the end of the growing season, whereas the default variety simulated a final peak in those traits. Both traits were further reduced by decreasing the fraction of remaining dry matter allocated to leaves, thus ensuring that leaf biomass was reduced and grain production increased. Both modifications had no influence on daily leaf number development, daily leaf number potential and final leaf number as well as daily stress potential for leaf area. In addition, growth stages were equal for the default and the modified variety, and there were no differences in modeled water demand or supply.

Simulation of the intercropping experiment with APSIM

Linking the relative intercropping system's components with the CANOPY module showed that APSIM was not able to simulate crop performance adequately. Those plants within the intercropping system – wheat and fieldpea – which were sown earlier than the companion species, maize, behaved according to the motto "the winner takes all" and were simulated adequately (fieldpea) or at least with a satisfactory bias (wheat) (Figure 4).

In the year 2008, the observed fieldpea dry matter yield at flowering was 3121 kg ha^{-1}, the simulated value 3145 kg ha^{-1}. Observed grain yield at harvest was 4121 kg ha^{-1}, 4731 kg ha^{-1} was simulated. At maturity 2008, wheat had 24506 kg ha^{-1} dry matter and 12400 kg ha^{-1} grain yield. The model simulated 21593 kg ha^{-1} and 10311 kg ha^{-1} respectively. In both intercropping systems, maize growing was not simulated or only implied. Maize was unable to develop grain in the wheat-maize system or fieldpea-maize system. In conformity with the competition module properties the simulation partitioned to the intercropped maize only 0.2% of the radiation fraction during the growing season on average. This drastically reduced radiation fraction disables maize to grow and develop.

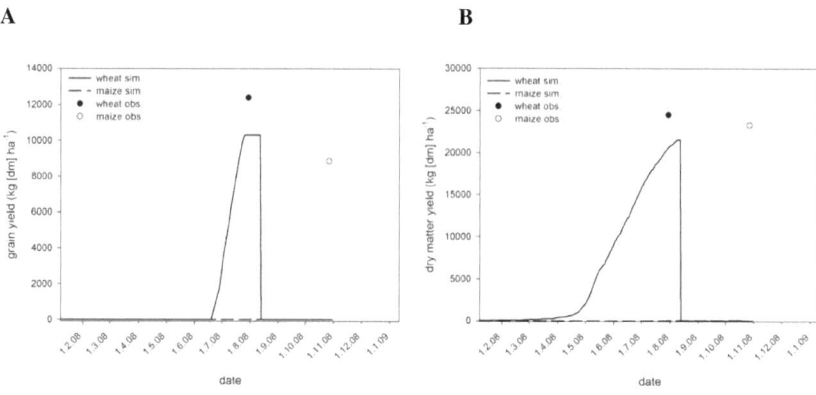

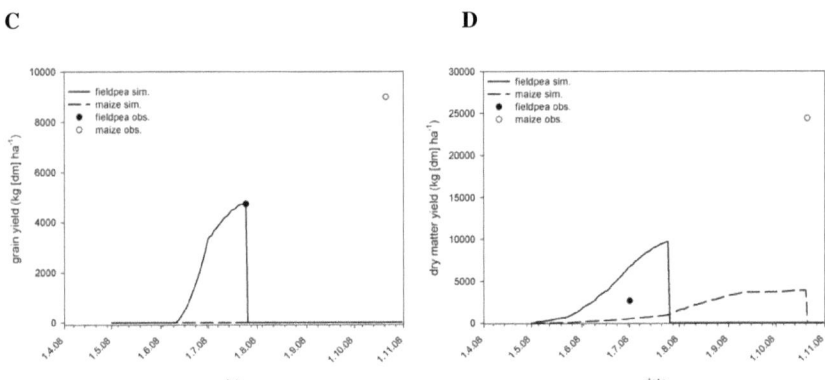

Figure 4: Simulated and measured grain (A/C) and dry matter yield (B/D) for maize/wheat (A/B) and maize/fieldpea (C/D) intercropping systems in the year 2008. Simulation was done using the APSIM crop growth model

This underestimation of maize growth occurs because the canopy module within the model was built to simulate mixed cropping and mixed canopy systems where competing species share a common soil water and nutrient pool and each plant species or crop row is located next to a different plant species or crop row. In addition, most mixed cropping systems are assumed to have a similar sowing date and a comparable plant height when development starts. The model could not deal with the development advance of one crop occurring in the relay intercropping of wheat and maize. By dividing the canopies into canopy layers, the understorey species gets scarcely any light because in the second layer (both canopies present), incoming light is reduced through the first layer (only taller plant canopy present) and, in addition, the species have to compete with each other. However, that approach does not reflect the reality of relay intercropping. The approach adopted in APSIM bares little resemblance to a relay intercropping system and its inadequacy in modeling the system is not surprising. In these systems where the competition effects occur in a few rows and the development of crops are staggered, plant growth cannot be modeled and simulated with the competition approach focused around the philosophy of Beer's law and represented in the APSIM CANOPY module. Point based models partition light, nutrients and water into pools which always ensure that the mass balance of the total system remains constant, and cannot model a system where this assumption is violated. If beneficial effects such as possible increases in light interception and possible changes in evaporation occur over multiple rows, then the total amount of light increases, and this cannot be represented in the current framework.

Increasing the yield of one species within intercropping systems without taking account of the performance of the companion species could not be realized with the modified Beer's law as competition algorithm.

Modeling competition for solar radiation and intercepted light is strongly based on simulated plant height as height defines the layers and the layers in turn define the fraction of light transmitted and entering the next layer below. In addition, LAI is distributed with height in the canopy. Total radiation intercepted in each layer is calculated according to the radiation transmission coefficient multiplied by LAI. If one species is much taller then the other, e.g. wheat in relay intercropping systems with maize, the maize canopy is assumed to be the lowest and intercepts barely enough light for plant growth. Modeled and observed plant height is shown in Figure 5 and shows the difficulties of taking simulated plant height as a basic premise. At maize sowing date, the wheat already has a height of about 50 cm and maize finds itself in the bottom layer. The module assumes that each maize row is surrounded by a wheat row and has to compete directly for light ignoring the fact that in strip intercropping there could be several rows of maize with row distances of 70 cm. Competition for solar radiation in those systems is reduced in comparison to mixed or row intercropping.

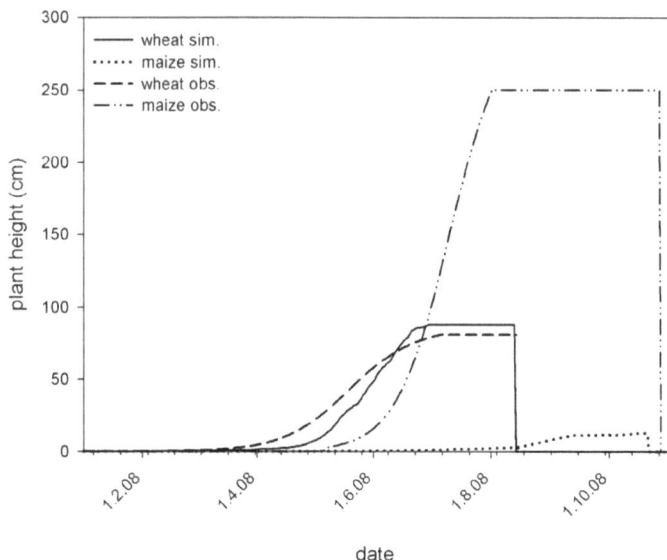

Figure 5: Simulated and observed maize and wheat height in the year 2008

DISCUSSION

After a re-evaluation of the APSIM model through minimal modifications of crop coefficients and variables such as biomass partitioning, senescence and/or thermal time calculation or thermal time unit requirements, APSIM was able to simulate dry matter and grain yield of German maize, winter wheat and fieldpea varieties grown under monocropping conditions adequately. The ability of APSIM - predominantly used and applied for simulating crop productivity in Africa and Australia - to simulate wheat performance in temperate European agro-ecological zones has already been shown by Asseng et al. [2] for the Netherlands. However, the APSIM Nwheat model used by Asseng et al. differed from the APSIM Vs 7.1 used in this study. The similarities between APSIM and DSSAT supported model adjustment as similar data requirements for both models were needed and the collected data for the evaluation of the DSSAT model could be used as well as DSSAT coefficients per se.

So far, the APSIM canopy module has been used for multiple cropping systems assuming that competing canopies were well mixed in the horizontal dimension. That may be the case for mixed cropping as well as row intercropping systems. Indeed, studies on wheat/lucerne and canola/lucerne companion farming [28], maize/cowpea intercropping [6][26] and maize/legume pasture and sorghum/legume pasture [6] intercropping indicated a good model fit between measured and observed yields and showed the current overall ability to simulate interspecific competition. Additionally, the APSIM model approach has proved to be a useful tool to address competition between crops and weed and the resulting yield loss [29]. The model was tested over a range of locations, seasons, sowing times, cultivars, plant densities, water regimes and N fertilizer rates.

Nevertheless, there has to be a clear definition and demarcation within multiple cropping systems, because each of those cropping systems is a system in its own rights with its own competition effects and characteristics. Models simulating strip intercropping might not be able to simulate mixtures and vice versa. It is important to give a brief definition of the outlook of the cropping system or the experiment, without becoming indistinct. Mixed cropping and intercropping are often used as synonyms, but when modeling mixed or intercropped systems, the design and sowing pattern becomes important. The (row) distance between competing species is essential when scaling competition or choosing competition factors. In mixed cropping, roots intermingle and interact. Plants assimilate nutrients from the same soil or nutrient pool, and the impact of neighboring plant height might be significant on the target plant. Not only plant height or LAI, but also individual species' compensation mechanisms, morphology or the appropriate timing of crop establishment of companion plants, govern competition.

So far, APSIM has been used to simulate mixed cropping or eventually row intercropping systems. Carberry et al. [6] suggested that the issue of spatially heterogeneous cropping systems also needs to be considered and that requires a different approach to the one currently implemented in APSIM. APSIM does not have the ability to model strip intercropping. Unless there is an issue of trade-off between the loss of crop yield in the intercrop relative to the sole crop [26], APSIM is able to address those losses due to competition for solar radiation. As soon as there is a winner, meaning a yield increase of the dominant intercropped species, without a loser, meaning an equivalent yield of the intercropped understorey species in comparison to the monocropped species, the modeling of competition for solar radiation using Beer's law is insufficient or not applicable. This approach does not allow for compensation growth of the understorey crop nor for yield increase of the dominant crop, as occurs in the relay intercropping of wheat and maize and the strip intercropping of fieldpea and maize. Competition for solar radiation in those systems is an important factor for crop productivity but neither the most dominant nor the one and only. In fact, to limit competition effects to competition for solar radiation only fails to accommodate the fact that other competition effects – such as changed nutrient availability or supply, soil chemistry or microclimate - occur and are of major importance [20][24].

In contrast to the APSIM model approach, the competition approach presented within the DSSAT model seemed to be a promising way to address more than one competition effect and showed that increasing solar radiation within the wheat strips explained or reflected only half of the yield increase [15]. The evaluated shading algorithm [16] allows for compensation growth of maize after wheat harvest as well as for increased wheat productivity. Introducing shade into modeling intercropping and relating its intensity or proportion to monocropping is a promising way to overcome the problem of different distribution of incoming sunlight within different distances from strip borders. In addition, increased sunlight for border rows in comparison to centered rows can be taken into account. Shade length can be calculated at each point of time during the day as well as each location in the world by means of longitude, altitude and sun azimuth; 100 % or near 100 % shading cannot occur because plant porosity and the distances between two competing plant rows will always allow some light to penetrate. In a similar context, the degree or proportion of shading was already calculated and used successfully. Stilma et al. [31] developed a minimal reference model for the population dynamics of annual weeds underneath a shading crop canopy by including treatment-specific shade functions. The questions are: Does the modeling of interspecific competition need 3D model frameworks or individual-based neighborhood models? And is each intercropping system too individual and are competition factors too diverse to be captured within one model?

As an example, there is a black box 'soil' to be considered. Microclimate in intercropping in comparison to monocropping could change and differ leading to an increasing air humidity or soil moisture [22][32]. Shading could lead to decreased soil temperature and vice versa [16]. Those microclimatic effects and differences could contribute to different soil properties or microclimate for soil microbes within intercropping in comparison to monocropping. Those effects are difficult to handle in modeling approaches. In their review Berger et al. [3] identified three major shortcomings in the modeling of competition among plants, which were the effects of plants on their local environment, adaptive behavior and below-ground competition. Neither of those could be addressed fully in current modeling approaches as local interactions and adaptational behavior to biotic and abiotic environment and stress can hardly be captured at plant population level [3].

In conclusion, modeling of intercropping with DSSAT offers a promising and completely different alternative to customary model approaches. However it requires considerable further improvements and in particular further validation. Modeling of intercropping with APSIM also needs to be significantly revised before it can be used for simulating strip or relay intercropping scenarios.

ACKNOWLEDGEMENTS

The authors' research topic is embedded in the International Research Training Group of the University of Hohenheim and China Agricultural University, entitled "Modeling Material Flows and Production Systems for Sustainable Resource Use in the North China Plain". We thank the German Research Foundation (DFG) and the Ministry of Education (MOE) of the People's Republic of China for their financial support. We also would like to thank Mrs. Nicole Gaudet for her diligent proof reading of the manuscript.

REFERENCES

[1]   Asseng, S., Keating, B.A., Fillery, I.R.P., Gregory, P.J., Bowden, J.W., Turner, N.C., Palta, J.A-, and Abrecht, D.G. (1998): Performance of the APSIM-wheat model in Western Australia. Field Crops Research 57, pp. 163-179.

[2]   Asseng, S., van Keulen, H., and Stol, W. (2000): Performance and application of the APSIM Nwheat model in the Netherlands. European Journal of Agronomy 12, pp. 37-54.

[3]   Berger, U., Piou, C., Schiffers, K., and Grimm, V. (2008): Competition among plants: Concepts, individual-based modelling approaches, and a proposal for a future research strategy. Perspectives in Plant Ecology, Evolution and Systematics 9, pp. 121-135.

[4]   CANOPY (2010): The APSIM canopy module. APSIM Vs 6 documentation files "Program_Files\Apsim6\apsim\canopy\docs\Canopy_science.htm".

[5]   Caldwell, R.M. (1995): Simulation models for intercropping systems. H. Sinoquet and P. Cruz (Eds.): Ecophysiology of Tropical Intercropping, Paris: INRA editions, pp. 353-368.

[6] Carberry, P.S., Adiku, S.G.K., McCown, R.L., and Keating, B.A. (1996): Application of the APSIM cropping systems model to intercropping systems. In: O. Ito, C. Johansen, J.J. Adu-Gyamfi, K. Katayama, J.V.D.K. Kumar Rao, and T.J. Rego (Eds.): Dynamics of roots and nitrogen in cropping systems of the semi-arid tropics. Japan International Research Center for Agricultural Sciences, pp. 637-648.

[7] Herndl, M. (2008): Use of modeling to characterize phenology and associated traits among wheat cultivars. Dissertation, Fakultät Agrarwissenschaften der Universität Hohenheim.

[8] Holzworth, D. and Huth, N. (2009): Reflection + XML simplifies development of the APSIM generic PLANT model. 18th World IMACS / MODSIM Congress, Cairns, Australia 13-17 July 2009, http://mssanz.org.au/modsim09.

[9] Inman-Bamber, N.G., Everingham, Y., and Muchow, R.C. (2001): Modelling water stress response in sugarcane: validation and application of the APSIM-Sugarcane model. 10th Australian Agronomy Conference, Hobart, Tasmania.

[10] IUSS Working Group (WRB) (2007): World Reference Base for Soil Resources 2006, first update 2007. World Soil Resources Reports No. 103. FAO, Rome.

[11] Jones, J.W., Hoogenboom, G., Porter, C.H., Boote, K.J., Batchelor, W.D., Hunt, L.A., Wilkens, P.W., Singh, U., Gijsman, A.T., and Ritchie, J.T. (2003): The DSSAT cropping system model. Europ. J. Agronomy 18, pp. 235-265.

[12] Keating, B.A. and Carberry, P.S. (1993): Resource capture and use in intercropping: solar radiation. Field Crops Research 34, pp. 273–301.

[13] Keating, B.A. and Meinke, H. (1998): Assessing exceptional drought with a cropping systems simulator: a case study for grain production in the north-east Australia. Agricultural Systems 57, pp. 315-332.

[14] Knörzer, H., Graeff-Hönninger, S., Guo, B., Wang, P., and Claupein, W. (2009): The rediscovery of intercropping in China: a traditional cropping system for future Chinese agriculture. E. Lichtfouse (Ed.): Springer Series: Sustainable Agriculture Reviews 2: Climate Change, Intercropping, Pest Control and Beneficial Microorganisms, Springer Science+Business Media, Berlin, pp. 13-44.

[15] Knörzer, H., Graeff-Hönninger, S., Müller, B.U., Piepho, H.-P., and Claupein, W. (2010): A modeling approach to simulate effects of intercropping and interspecific competition in arable crops. International Journal of Information Systems and Social Change 1, pp. 44-65.

[16] Knörzer, H., Grözinger, H., Graeff-Hönninger, S., Hartung, K., Piepho, H.-P., and Claupein, W. (2011): Integrating a simple shading algorithm into CERES-wheat and CERES-maize with particular regard to a changing microclimate within a relay-intercropping system. Field Crops Research (in press).

[17] McCown, R.L., Hammer, G.L., Hargreaves, J.N.G., Holzworth, D.P., and Freebairn, D.M. (1996): APSIM: a novel software system for model development, model testing and simulation in agricultural systems research. Agricultural Systems 50, pp. 255-271.

[18] McCown, R.L., Hammer, G.L., Hargreaves, J.N.G., Holzworth, D.P., and Huth, N.I. (1995): APSIM: an agricultural production system simulation for operational research. Mathematics and Computers in Simulation 39, pp. 225-231.

[19] Meier, U. (1997): Entwicklungsstadien Pflanzen (BBCH Monograph). Biologische Bundesanstalt für Land- und Forstwissenschaft (Ed.), Blackwell Wissenschafts-Verlag, Berlin-Wien [u.a.].

[20] Meinke, H., Carberry, P.S., Cleugh, H.A., Poulton, P.L., and Hargreaves, J.N.G. (2002): Modelling crop growth and yield under the environmental changes induced by windbreaks. 1. Model development and validation. Australian Journal of Experimental Agriculture 42, pp. 875-885.

[21] Monsi, M. and Saeki, T. (1953): Über den Lichtfaktor in den Pflanzengesellschaften und seine Bedeutung für die Stoffproduktion. Jpn.J.Bot. 14, pp. 22-52.

[22] Njoku, S.C., Muoneke, C.O., Okpara, D.A., and Agbo, F.M.O. (2007): Effect of intercropping varieties of sweet potato and okra in an ultisol of southwestern Nigeria. African Journal of Biotechnology 6, pp. 1650-1654.

[23] Probert, M.E., Keating, B.A., Thompson, J.P., and Parton, W.J. (1995): Modelling water, nitrogen, and crop yield for a long-term fallow management experiment. Aust. J. Exp. Agric. 35, pp. 941-950.

[24] Raynaud, X. and Leadley, P.W. (2005): Symmetry of belowground competition in a spatially explicit model of nutrient competition. Ecological Modelling 189, pp. 447-453.

[25] Ritchie, J.T. (1972): Model for predicting evaporation from row crops with incomplete cover. Water Resour. Res. 8, pp. 1204–1213.

[26] Robertson, M.J., Carberry, P.S., Wright, G.C., and Singh, D.P. (1997): Using models to assess the value of traits of food legumes from a cropping systems perspective. Invited paper for the International Food Legumes Conference, September 1997, Adelaide.

[27] Robertson, M.F., Cawthray, S., Birch, C.J., Bidstrup, R., and Hammer, G.L. (2003): Improving the confidence of dryland maize growers in the northern region by developing low-risk production systems; Solutions for a better environment. Proceedings of the 11th Australian Agronomy Conference Geelong, Victoria.

[28] Robertson, M., Gaydon, D., Latta, R., Peoples, M., and Swan, A. (2004): Simulating lucerne/crop companion farming systems in Australia. "New directions for a diverse planet", Proceedings of the 4th International Crop Science Congress, 26 Sep – 1 Oct 2004, Brisbane, Australia. Published on CDROM, Web site www.regional.org.au/au/cs.

[29] Robertson, M.J., Whish, J., and Smith, F.P. (2001): Simulating competition between canola and wild radish. ARAB 2001.

[30] Spitters, C.J.T., Toussaint, H.A.J.M., and Goudriaan, J. (1986): Separating the diffuse and direct component of global radiation and its implications for modeling canopy photosynthesis, Part I: Components of incoming radiation. Agricultural and Forest Meteorology 38, pp. 217-229.

[31] Stilma, E., Keesman, K.J., and Van der Werf, W. (2009): Recruitment and attrition of associated plants under a shading crop canopy: Model selection and calibration. Ecological Modelling 220, pp. 1113-1125.

[32] Williams, P.A. and Gordon, A.M. (1995): Microclimate and soil moisture effects of three intercrops on the rows of a newly-planted intercropped plantation. Agroforestry Systems 29, pp. 285-302.

10 General Discussion

In the present thesis, the chapters I, II, III, V and VI are related, with chapter IV being an excursus into peanut production in China. The chapters deal either with intercropping or modeling intercropping or combine both aspects as shown in chapter II. Each chapter can be read independently and is discussed independently at its end. Thus, the aim of this general discussion is not to discuss the chapters in succession, but as an overall perspective of designing, modeling and evaluating intercropping systems with special regard to the North China Plain. Aspects and features introduced in the beginning of this thesis will be re-iterated, and the subtleties between intercropping and modeling intercropping issues will be studied critically.

Intercropping systems are the epicenter between traditional farming systems and modern production modes, and will be evaluated according to their performance, their contribution to sustainability and production, and to their potential for future agriculture. Modeling intercropping is a challenge. Although it is studied across a wide range of situations only a few issues have so far been investigated. Competition effects like solar radiation are difficult to handle system spanning. If there are black dots on the landscape of intercropping, there is a black box 'soil' for modeling competition.

THE FUTURE OF INTERCROPPING?

Without doubt intercropping is a cropping system of research interest. According to the keyword request 'intercropping' within the Scopus database, more than 2 800 topic related papers and articles can be found. Across many disciplines, e.g. plant nutrition (Inal et al., 2007; Zhang et al., 2004), crop production (Chen et al., 2004; Ghaffarzadeh et al., 1994; Ayisi et al., 1997; Lesoing and Francis, 1999; Pridham and Entz, 2008), agroecology (Jolliffe, 1997; Ren, 2005; Sanders, 2000), entomology (Ma et al., 2007), modeling (Caldwell, 1995; Keating and Carberry, 1993; Knörzer et al., 2010) as well as cropping systems studies, various studies dealt with issues related to intercropping in order to investigate competition effects, species interaction, optimal species combination, optimization of cropping systems, intercropping benefits, sustainability, agricultural practice and soil and plant properties changes and exchanges. Various intercropping systems are widespread all over the world, especially in Asian and African countries, but also in Latin America, Europe and the United States of America. Multiple cropping systems seem to be as old as the history of agriculture. The question is does intercropping have the potential for next generation agriculture in countries where **traditional farming systems** and structures will turn into **modern**

production modes with mass production, high yield, mechanization, and high inputs being on headlines?

During the last couple of decades, China's agriculture has turned more and more from subsistence farming to a nation's self supplier, and an exporter of agricultural goods. The export value of food and live animals used chiefly for food (in million US dollars) tripled from 1999 to 2009 (China Yearly Macro-Economics, 2010). The fast development of the structural agricultural sector, mechanization and production growth is still ongoing. Within ten years, from 1999 to 2009, the gross output value of farming in Henan province rose from 14 106 to 30 611 million Yuan (China Yearly Macro-Economics, 2010). China belongs to those countries where intercropping is widespread on the one hand and modern production modes are developing on the other. Chapter I within this thesis deals with the distribution, characteristics and benefits of Chinese cereal intercropping systems in detail.

According to Li (2001) and Li et al. (2007), 20 to 25 % of China's arable land is under intercropping. Thus, intercropping seems to be a big issue in China. Having a closer look at major cereal production provinces like Hebei, a trend can be detected. Whereas in former times a relay intercropping of wheat and maize was predominant, nowadays plant breeding and mechanization contributed to a decline of that system that is being replaced by a double cropping system without overlapping growing season of both species (Wang et al., 2009). Fast maturing maize varieties and, in addition, the use of machines for wheat harvest instead of labor and time intensive cutting by hand were the reasons for replacing the traditional relay intercropping by a double cropping system. Within the last few decades, the degree of mechanization and the numbers of machines has increased steadily. Since 1980 the use and the application rate of chemical fertilizer and pesticides have been increased explicitly. The total power of agricultural machinery doubled and the consumption of fertilizers increased 2.5 fold between 1998 and 2008 (China Yearly Macro-Economics, 2010). China's cropping systems turned from low input systems to high input systems. Intercropping is known as an adjusted cropping system for low input but to lose ground when applying high input. The apparent unlimited host of labor force in rural China is declining too, as more and more rural workers move into cities in order to increase their income. In addition, rural household income is increasingly derived from non-farming work (internal IRTG presentations). As a result and as a future trend scenario, it is expected that through increasing mechanization, agriculture will be more efficient in terms of labor demand, full-time farmers will enlarge their farm size, and crop production will be intensified. Intercropping is known as labor and time intensive, feasible on small-scale farming, and profitable when low input driven. Hence, does the agricultural development oppose the traditional persistence of intercropping systems?

10. General Discussion

The disappearing wheat-maize relay intercropping system in the North China Plain is an example of the decline of intercropping practice. It could be opposed that new intercropping systems evolved in China too. In the 1980's the wheat-peanut relay intercropping was established and promoted, especially through the Shandong Peanut Research Institute (SPRI) (Gowda et al., 1996). This might be ideal as intercropping per se might not be replaced by monocropping or double cropping systems but could change, alter or diversify in the course of time. It is not static, in contrast, within intercropping's variability, adaptability and flexibility lies its potential for future persistence. To be provocative: intercropping has not only the status quo of being still alive in fast developing countries, but it emerges also in industrialized countries. As chapter I is entitled, intercropping is experiencing some kind of rediscovery. There is an increasing awareness deduced from organic farming, agroecology, and the debate around sustainability. To maintain or further improve intercropping systems, they have to be accepted by the farmer, because any agricultural practice must provide advantages over other available options in the eyes of the practitioner (Sullivan, 2003).

For Tom Frantzen, a farmer in Iowa (USA), the advantages of strip intercropping are that the strips can be considered as a crop rotation within one big field (Sullivan, 2003) as well as its economical benefits which are about $19 per acre higher for the strip intercropping than for the monocropped equivalent. Increased pest suppression, soil building advantages, and yield increase are the reasons for agronomist R. Cruse from the Iowa State University (Cruse, 1996) to encourage farmer taking intercropping into consideration (Sullivan, 2003). In comparison to scientific research, farmers have to be much more pragmatic concerning what sustainable agriculture means. Paul Mugge, a 320 acre farmer in Sutherland (USA), is practicing strip intercropping, because he and his wife want to obtain the most net profit from each acre and each hog (Kendall, 1996/1997). His long term vision is to be profitable, to be efficient in the terms of resources, to understand more about ecology and to use that understanding. He says (Kendall, 1996/1997): "I want my farm to contribute more than its share to feeding the world while contributing much less that its share to environmental degradation." Marvin J. Williams Jr. from Michigan (2003) possesses a United States Patent of an intercropping system. While explaining his invention's aims, he emphasizes the production of commercial plants for commercial machine-driven agricultural methods. Avoidance of pesticides and herbicides, effective ground cover, soil enrichment, and increased drought resistance were the decision making reasons for him to practice intercropping.

Strip intercropping is often named as **'ecologically innovative agriculture'** in the USA, this coupled with the above American farmer examples indicates that intercropping is not limited to traditional and old-fashioned smallholder farming, remote areas, or developing countries. Even if

10. General Discussion

intercropping systems tend to disappear or decline when there is development from small scale and subsistence farming to farming on bigger scales, mostly connected to monocropped fields and where the arable land area is feasible for large, plain fields like in the North China Plain, there is a potential for intercropping that should not be neglected. The future of intercropping, viewed globally, might be that two doors will close, but one will open. Modern agricultural practices and especially the agro-industrialized cultivation have to be reflected in some critical way, as the claims and demands upon agricultural production raise and change. Besides food security, sustainability, ecological and environmental friendly production modes, healthy food, and conservation will be issues agriculture and agricultural politics have to deal with. For example, the Büro für Technikfolgen-Abschätzung beim Deutschen Bundestag (TAB) in its TAB letter no. 29 (2006) designated mixed intercropping as an ecological and economical potentially suitable 'modern agricultural engineering and production method'. The TAB advises the German parliament on changes in technical and social matters.

While achieving high yields, increasing labor efficiency or increasing mechanization remain major objectives there is no way there will be a back-to-the roots development. Agro-romantic visions are inappropriate when dealing with global food safety or income security for peasants in developing countries, remote areas, or facing social disadvantages.

Reproaches generated against polycultures, multiple cropping systems, or intercropping were that they lack the potential of producing meaningful marketable surplus, that indigenous knowledge will not yield panaceas for agricultural problems in the world (Altieri, 2000), that **intensification of production** is essential for the transition from subsistence to commercial production (Blauert and Zadek, 1998), and that intercropping relies on hand tools and draft animals and cannot be mechanized in a large scale. US farmers like Tom Frantzen, Paul Mugge or Marvin J. Williams Jr. (Sullivan, 2003) showed that indeed, intercropping could be an economical as well as ecological alternative way of crop production. To handle intercropping as an option for production, as a production branch, and not as the one and only way offers the possibility of turning a cropping system to account for averting risks, saving inputs, increasing biodiversity, decreasing soil erosion, and simultaneously, stabilize and secure income. Innis (1997) showed in his overview over studies calculated the LER of inter- in comparison to monocropping systems that proportional yields or the output per unit are higher in intercropping systems than in monocropping. In addition, Altieri (2000) stated for the 1980s that in Latin America peasant polycultures' overall contribution to the general food supply reached approximately 41 % of the agricultural output for domestic consumption. Monocropping does not at all guarantee high yields. Besides, the focusing on partial aspects of single crops rather than total system yields of multiple crops as well as the focusing on quantity per acre rather than nutrition per acre neglects the fact that multiple cropping systems generally enhance total farm productivity rather than the yield of specific components (Altieri, 2003). Maybe intercropping has a need of a different rating of criteria than monocropping to evaluate its contribution to global agriculture and food supply.

At least, the claim that intercropping could not be **mechanized** and thus, it would be rather difficult to integrate it into modern monocropping systems is questionable. Exemplary, the US farmers are to be mentioned again. To apply regular herbicide treatments, Tod Intermill from South Dakota just uses a ground sprayer of strip width (Sullivan, 2003). Sullivan (2003) also reports from a US farmer who uses a 12-row planter for a maize-soybean intercropping system consisting out of six rows of maize and six rows of soybean. To establish the six-row strips he fills the middle six hoppers with maize and the outer three hoppers with soybean. Lloyd Younger (1978, [57]) from Illinois owns an United States patent for an apparatus for sowing a second crop in a standing crop in order to mechanize relay intercropping (Figure 6): Seeding of the crop is performed using a self-driven, preferably three-wheel vehicle carrying a grain drill box having flexible tubes which fit between the rows of grain and each of which discharges the seed into a gap between two downward-inward slanted discs which first cut a slit in the ground, then deposit the seed and finally cover the slit. Obviously, there are ways and methods to mechanize strip or relay intercropping. So far, there is no commercial production of those machines, which might be related to the fact that there is no big demand. In countries where intercropping is the most widespread, farmers do not have the capital to invest into machines. Nevertheless, the mechanization and productivity of intercropping is not impossible but it appears like the causes are rather social than technical (Altieri, 2003).

To intensify or to preserve the various intercropping systems in China and all over the world will be a challenge worth taking and a task for further research. Every system seems to be a system on its own and worth to be documented and/or analyzed. Although there have been a lot of studies about intercropping so far, there are still uncertainties, **black spots** and open questions. To study competition effects more closely and to test new species combinations, to optimize input amounts

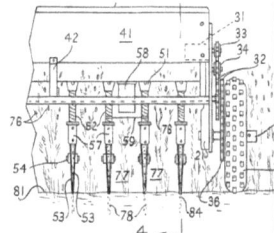

Figure 6: Fragmentary rear elevational view of a portion of an apparatus for sowing a second crop in a standing crop (left) (Younger, 1978). Strip intercropping appearance and practice in the USA (right) (http://www.thisland.illinois.edu/60ways/60ways_17.html)

and applications, and to evaluate or breed suitable intercropping varieties in order to improve or adjust existing intercropping systems might be a more consistent approach for stimulating agricultural production than approaches undertaken in the first Green Revolution.

THE FUTURE OF MODELING INTERCROPPING?

A possibility to improve and adjust intercropping systems, and to pick up on questions which arose in the introduction of this thesis, is to model and simulate cropping systems. In chapter II, an overview was given over various models which already dealt with intercropping or interspecific competition. Whereas competition about solar radiation was the most common and advanced competition factor studied and detected, other factors were regarded, too, or at least seem to be of great influence. Examples are: plant density, self-thinning, and mortality in a plant population (Yokozawa and Hara, 1992), belowground competition or competition for water (O'Callaghan et al., 1994; Ozier-Lafontaine et al., 1995, 1998; Raynaud and Leadley, 2005), and competition for soil nutrients (Corre-Hellou et al., 2009; Hauggaard-Nielsen et al., 2006; Ibeawuchi, 2007; Jensen, 1996).

Competition for solar radiation, shading, and light interception are one of the key factors within multiple cropping systems, but not the one and only or of such importance that other factors could be neglected. Explicitly, the term 'multiple cropping' instead of 'intercropping systems' was used here, as multiple cropping systems like sequential cropping, mixed cropping, row or strip intercropping, and relay intercropping (Federer, 1993) differ greatly considering competition factors, competition impact and the ability to compensate competition. Models simulating strip intercropping might not be able to simulate mixtures as well and vice versa. In this case, it is important to give a brief definition about the outlook of the cropping system or the experiments, without becoming indistinct. Otherwise, the studies are entitled to analyze or simulate intercropping, but the cropping system is a mixture (Carberry et al., 1996; Hauggaard-Nielsen et al., 2006). Mixed cropping or intercropping are often used as synonyms, but when modeling mixed or intercropped systems, the design and sowing pattern becomes important.

Models have to reach the point from evaluation and validation to application. As reviewed in chapter II, a lot of studies about intercropping and interspecific competition exist and are published, but the question is, to which extent they are applicable for the use in the field or for making recommendations for farmers to improve there cropping strategies? Otherwise, are the models more applicative for modeling mixed cropping, row intercropping or the modeling of weed-cereal interactions than for strip or relay intercropping? Are the models over parameterized or is each system too individual and the competition factors too diverse to be captured within one model?

Some or most of the models used for modeling competition about solar radiation within intercropping or mixed cropping are based upon **Beer's law** or a modification of Beer's law (Carberry et al., 1996; Corre-Hellou et al., 2009; Debaeke et al., 1997; Sonohat et al., 2002; Tsubo et al., 2005). Beer's law is defined as the linear relationship between absorbance and the concentration of an absorber of solar radiation, whereas absorber concentration equals canopy layers when Beer's law is assigned to plant growth models. Light transmission is then computed as a negative exponential function of the downward cumulated leaf area index (LAI) and the plant extinction coefficient of each canopy. This approach has proven to be sufficient and successful in simulating mixed cropping, weed-cereal interactions, row intercropping, and additive intercropping systems. Nevertheless, in chapter VI it was shown, that for modeling strip or relay intercropping or a replacement intercropping system, the approach is not suitable. Within those cropping systems and according to that approach, the dominant species behave like the winner takes all. Point based models partition light, nutrients and water into pools that always ensure that the mass balance of the total system remains constant, and cannot model a system where this assumption is violated. If beneficial effects, like possible increases in light interception and possible changes in evaporation occur over multiple rows then the total amount of light increases and this cannot be represented.

The (row) distance between competing species is essential when scaling competition or choosing competition factors. In mixed cropping, roots intermingle and interact. Plants assimilate nutrients from the same soil or nutrient pool, and the impact of neighboring plant height might be significant on the target plant. The Beer's law approach ignores the possibility of different LAI distributions within a layer (CANOPY, 2010) (uniform distribution), and instead distributes LAI with height in the canopy. Thus, it is assumed that 47 % of the leaf area is located in the top 10 % of plant height, 27 % in the next 10 %, 15 % in the next 10 %, and so on. This assumption might be appropriate for monocropping or mixed cropping, but within strip intercropping or relay intercropping, where neighboring species hold increased individual space at the strip borders because of development advance (relay intercropping) or row distance (replacement intercropping systems, strip intercropping), circumstances change. As a result, the understorey species, located in the second layer according to the Beer's law approach, might receive more light than assumed or can compensate for the decreased incoming solar radiation after the dominant species is harvested. Compensation growth or mechanisms cannot be respected. Then, neighboring plant height and LAI are less determining for competition. Russell et al. (1989) confirmed that a non-uniform distribution of leaf area increases both the amount of PAR absorbed by the canopy and the amount of photosynthesis by comparison with a uniform distribution of leaf area.

Not only plant height or LAI govern competition, but individual species compensation mechanisms, morphology or the appropriate timing of crop establishing of companion plants (Tsubo and Walker, 2004), too. In fact, the restriction for competition for solar radiation neglects the fact that other competition effects - like changed nutrient availability or supply, soil chemistry, or microclimate - occur and are of major importance (Meinke et al., 2002; Raynaud and Leadley, 2005).

There is a **black box 'soil'** to be considered. Two aspects influence soil related processes which in turn influences intercropping systems performance. First, a changing microclimate in intercropping in comparison to monocropping could increase air humidity or soil moisture (Njoku et al., 2007; Williams and Gordon, 1995), as well as shading could lead to decreased soil temperature and vice versa (Knörzer et al., 2010) (chapter V). Those effects could contribute to different soil properties or microclimate for soil microbes within intercropping in comparison to monocropping. Those effects are difficult to handle in modeling approaches if climate features are taken from standard weather stations without taking microclimate within the field into consideration. Mass balance of the total system remains still constant, but some focuses shifted and have to be re-estimated. Taking microclimate into consideration while modeling intercropping, as performed in a first modeling approach in chapter V, could improve the simulation excellence.

Secondly, initial soil properties as well as species combination could induce processes different to those within monocropping. Raynaud and Leadley (2005) found that nutrient uptake which is not proportional to size differences could occur in soils with higher water content when plants differ in nutrient uptake kinetics, and the mechanisms leading to size-asymmetry are very different from light competition since they arise from physiological differences in resource uptake capacity. In addition, Jensen (1996) stated that within a pea-barley intercropping system, the proportion of total N derived from fixation in the intercrop was significantly increased compared to the total N derived from fixation in monocropped pea. However, a rule or standard algorithm defining that N fixation is increased if legumes and cereals are combined, cannot be determined or found simplistic. Hauggaard-Nielsen et al. (2006) showed for their pea-barley intercropping experiments with different plant densities, that only at low cropping densities intercropped pea increased its reliance on N_2 fixation relative to sole cropping.

Thus, modeling intercropping by setting the initial soil conditions equal to those of equivalent monocropping situations or setting one combined nutrient pool for companion crops assuming a daily alternating nutrient and water supply is not adequate even if yield or dry matter accumulation might be simulated appropriately. With this approach, competition cannot be simulated and the modeling is restricted to fit the plant growth curve only without having a knowledge gain throughout modeling. Berger et al. (2008) identified three major gaps within their review about modeling competition among plants, which were the effects of plants on their local environment, adaptive behavior, and below-ground competition. Neither of those could be addressed fully in current modeling approaches as local interactions and adaptational behavior to biotic and abiotic environment and stress can hardly be captured on plant population level (Berger et al., 2008).

The modeling approach in the present thesis (chapters II and V) differed from those discussed above. Microclimate influence was taken into consideration as well as a more empirical than mechanistic modeling was chosen by introducing a **shading** algorithm. Nevertheless, this approach can be converted into a mechanistic one by further studying attributes of shade (distribution and proportion of PAR, porosity of species for neighboring species, shading intensities etc.), which is needed for further model improvement. Introducing shade into modeling intercropping and relating its intensity or proportion to monocropping is a promising way to overcome the problem of different distribution of incoming sunlight within different distances from strip borders. In addition, increased sunlight for border rows in comparison to centered rows can be taken into account. Modeling intercropping with the evaluated shading algorithm within DSSAT is still an approach, but could be regarded as a model testing for the potential hidden in 'shadow'. It showed that competition for solar radiation is a major factor in modeling intercropping, but not the one and only. For the relay intercropping experiment with wheat and maize, the shading algorithm explained half of the yield increase in wheat (chapters II and V), but showed, too, that other competition effects are important to consider which lead away from the accentuation of modeling competition for light solely.

In addition, it has the advantages that it could be introduced into other existing models rather easy; shade length can be calculated at each point of time during the day as well as each location of the world by means of longitude, altitude, and sun azimuth; shading of approximately 100 % cannot occur because of plant porosity and distances from two competing plant rows, where light could penetrate, in comparison to Beer's law and the division of canopies into different layers where the later established understorey species gets hardly any light and thus, could not assimilate sufficient light for growing within simulation. The shading approach might be more cropping system spanning and global.

In other coherencies, the degree or proportion of shading was calculated and used successfully. Stilma et al. (2009) developed a minimal reference model for the population dynamics of annual weeds underneath a shading crop canopy by including treatment-specific shade functions. In addition, the SOMBRERO model from Niewienda and Heidt (1996) was used to simulate complex shadow sceneries and to estimate the influence of shadows on passive solar systems by calculating the quantities and time dependency of geometrical shadow coefficients (Yezioro and Shaviv, 1994) and the proportion of a shaded area as a function of time and location. They stated that the shading depends in many ways on the orientation of the collecting surface (which is an understorey canopy in the case of competing plant growth modeling), its surroundings (neighboring dominant plants), on the season and the time of day. The question is, if modeling of interspecific competition needs 3D model frameworks comparable to the SOMBRERO tool or individual-based neighborhood models like Berger et al. (2008) indicated?

It is obvious that there is a need for further analysis of shade, the introducing of shading algorithms and the testing whether it is necessary to link different modeling frameworks with existing ones. Nevertheless, to enable process-oriented plant growth models to simulate various intercropping scenarios, the shading approach seems to be promising.

11 Summary

Intercropping is a good deal more than the survival of the fittest. Defined as the cropping of two or more crops within the same or an overlapping growing season and within the same field, **intercropping** offers a great variation of species combination, benefits as well as challenges for cropping systems design and farmers. Carefully balanced between facilitation and competition, intercropping bears the potential of increased yield and yield stability, income security, resource use efficiency and biodiversity. The contribution of intercropping systems, mainly practiced in Asian and African countries, should not be underestimated or neglected when considering rural income with smallholder farming predominant. In less developed countries, remote areas and on degraded soils, intercropping contributes substantially to local marked supply, diet diversification and sustainability. Under certain circumstances, intercropping is a concept with substance.

Taking China as an example where intercropping is widespread and approximately one third of the arable land is under intercropping, the system has a 3 000 year old tradition, is as diversified as the Chinese history, reduces pesticides and fertilizer consumption while performing even better under low than high input, reduces nitrate leaching and soil erosion, reduces pests and diseases and produces more output in a more sustainable way than monocropping systems in agro-industrialized monocropping systems. Intercropping gives evidence about traditional cropping systems with the potential for future production systems under the paradigm of sustainability. Wherefore North-American and European farmers rediscover those systems as an option to fulfill issues of modern agricultural production modes.

High yield and sustainability are the catchphrases of production in the 21st century – and agronomy research has to provide solutions in increasingly briefer terms. Thus, **improved cropping strategies** have to be developed. Taking modeling and simulation tools into account could help both to accelerate research attainments and to give a better understanding of cropping systems. How intercropping systems have to be designed and what problems arose, how datasets have to be evaluated for modeling and what kind of multi-level interactions has to be taken into account are the basic topics within the present thesis. The thesis has been carried out within the Sino-German International Research Training Group "Modeling Material Flows and Production Systems for Sustainable Resource Use in intensified Crop Production in the North China Plain", and was financed by the German Research Foundation and the Chinese Ministry of Education. In this context, field experiments were conducted with a maize/peanut strip intercropping system in China

11. Summary

and a maize/wheat as well as a maize/fieldpea strip intercropping system in Germany between the years 2007 and 2009.

The results indicated that border effects are the key component of intercropping performance, and modeling strip intercropping could be termed as modeling field borders. Nevertheless, analyzing strip intercropping has peculiarities as they lack in randomization as the cropping system imposes alternating strips. Thus, **spatial variability** and its effect on yield were regarded differently. For statistical analysis of the trials, different spatial models were applied to account for the spatial trend and to check whether or not standard models are suitable for analyzing strip intercropping experiments.

Throughout a review and evaluation of available data of cereal intercropping systems in China, four **intercropping regions** could be classified and distinguished. The four intercropping regions are the Northeast and North, the Northwest, the Yellow-Huai River Valley and the Southwest. Going from north to south, the cropping systems change from one crop a year with a great potential for intercropping to relay intercropping of especially maize and wheat and double cropping systems and at least three cropping seasons per year with different kinds of rotations and rotations replacing intercropping. The species spectrum within those systems is rather wide and includes wheat, maize, cotton, green manure, soybean, sweet potato, rape, peanut, broomcorn millet, bean, buck wheat, millet, tobacco, sorghum, rice, cassava, garlic and a great variety of vegetables.

In comparison to other countries, like India or Africa, intercropping systems in China seemed to be less documented and studied with the point of view to adjust or improve production patterns or to select for varieties specifically suitable for intercropping. Nevertheless, there is a vast amount of intercropping studies all over the world. In addition, there are a lot of studies about **modeling intercropping** or interspecific competition. The various approaches seem to be promising as the validation of the divers models showed. Nevertheless, most of the model studies restricted the competition effects to competition for solar radiation, not taking microclimate or soil effects into account. Although a lot of crop growth and weed models have been used to simulate intercropping and interspecific competition, only a few show completely different approaches or ideas how to simulate competition in general. Most often, a modified Beer's law approach was used.

An example for a process-oriented model that has been used for simulating mixed intercropping scenarios by linking sole crop models with a Beer's law implemented subroutine is the APSIM model. It was used as an example for a **model comparison** between a model already able to simulate interspecific competition and the DSSAT model, which was used predominately in this thesis to model the conducted strip intercropping experiments. So far, DSSAT was not able to model intercropping. Thus, a shading algorithm was evaluated and tested for DSSAT. In addition,

11. Summary

the impact of changing microclimate within strip intercropping was analyzed. The Beer's law approach was not capable to model strip intercropping. Unless there is an issue of trade-off between the loss of crop yield in the intercrop relative to the sole crop, those model approaches are able to address to those loss due to competition for solar radiation. As soon as there is a winner, meaning a yield increase of the dominant intercropped species, without a loser, meaning an equivalent yield of the intercropped understorey species in comparison to the monocropped species, the modeling of competition for solar radiation using Beer's law is insufficient or not applicable. This approach does not allow for compensation growth of the understorey crop neither for yield increase of the dominant crop. Competition for solar radiation in those system is a driving force for crop productivity but neither the most dominant nor the one and only.

In contrast, the model approach with DSSAT showed, when applying a simple **shading algorithm** that estimated the proportion of shading in comparison to the monocropping situation and in dependency from neighboring plant height both systems performance could be simulated adequately. No submodel was needed to be introduced into the model, but instead, the standard weather file was modified. The advantage of that method is that it could be adapted for all species combination, and a simpler approach could be an adequate surrogate for a complex coherence.

Besides shading or competition for solar radiation, changing **microclimate**, e.g. soil temperature and wind speed, as well as its influence over time has to be taken into regard when modeling intercropping. Resource distribution and allocation in space and time seems to be more important than the total amount of resources. The linear shading algorithm modeling approach showed, that modifying only the incoming solar radiation did not explain fully the yield increase. Foremost the changing of initial conditions taking the higher amount of N cycling in the system, because of the increased soil temperature and N mineralization into account could explain the increase. Increased or decreased wind speed in border rows compared to centered rows might change the transpiration rate as well as the amount of CO_2 assimilation. Those effects have to be taken into account when simulating interspecific competition.

11 Zusammenfassung

Intercropping ist mehr als nur ein „survival of the fittest". Definiert als der Anbau von zwei oder mehr Feldfrüchten auf der gleichen Fläche und innerhalb der gleichen oder einer sich überlappenden Vegetationsperiode, bietet **Intercropping** eine große Bandbreite an Kombinationsmöglichkeiten von Feldfrüchten, verbunden mit vorteilhaften und nachhaltigen Effekten für die jeweiligen Kulturarten. Intercropping ist aber gleichzeitig eine Herausforderung für jeden Landwirt und stellt hohe Ansprüche an die Gestaltung des jeweiligen Produktions- oder Anbausystems. Letztendlich ist Intercropping ein Balanceakt von mindestens zwei Kulturarten, die sich wechselseitig begünstigen oder um Wasser, Nährstoffe oder Sonnenlicht konkurrieren. Gelingt dem Landwirt mit der Wahl geeigneter Kulturarten und Anbausystemen dieser Balanceakt, hat Intercropping das Potential zum erfolgreichen System. Dessen Vorteile sind unter anderem höhere Erträge, Ertrags- und Einkommenssicherheit, effizientere Ressourcennutzung und höhere Biodiversität. Den Beitrag, den Intercropping vor allem in Afrika und Asien hinsichtlich Produktionsvolumen, Belieferung örtlicher Märkte und Nahrungsmittel Diversifizierung liefert, ist nicht zu unterschätzen oder zu vernachlässigen. Intercropping wird hauptsächlich in kleinparzellierten Agrarlandschaften, in kleinbäuerlich strukturierten Ländern Asiens und Afrikas und in benachteiligten Gebieten betrieben, wo es beträchtlich zum Einkommen der ländlichen Bevölkerung und zu einer nachhaltigen Produktion beiträgt. Somit ist Intercropping ein Anbausystem das einerseits ein hohes Potenzial bietet, andererseits aber auch sehr große Herausforderungen an den Landwirt stellt.

Ein Beispiel für ein Land, in dem Intercropping weit verbreitet ist und auf eine 3 000 Jahre alte Geschichte zurückblicken kann, ist China. Schätzungen zufolge wird Intercropping auf rund einem Drittel der gesamten Anbaufläche Chinas praktiziert. Dortige Anbauspektren und Produktionssysteme sind so vielfältig und abwechslungsreich wie die Geschichte des Landes selbst. Der Einsatz von Pestiziden und chemischen Düngemitteln kann reduziert, Krankheits- und Schädlingsbefall eingedämmt und Nitrat-Auswaschung und Bodenerosion vermindert werden. Intercropping gilt als ein Anbausystem, welches bei geringerem Betriebsmitteleinsatz höhere Erträge oder Gewinne erzielt, verglichen mit den ausgedehnten Monocropping Systemen moderner Agrar-Industriebetriebe. Damit belegt Intercropping, dass in traditionellen Anbausystemen ein Potential für zukünftige und nachhaltige Produktionssysteme schlummert. Nicht ohne Grund haben auch Landwirte in Nord-Amerika und Europa Intercropping wieder- oder neu für sich entdeckt, denn die Paradigmen für eine landwirtschaftliche Produktion haben sich im 21sten Jahrhundert

11. Zusammenfassung

geändert. Neben höheren Erträgen spielt die Art und Weise der Produktion, sprich deren Nachhaltigkeit und Ressourcenschonung, eine zunehmende Rolle.

Um diesen Paradigmen und um politischen, sozialen und ökonomischen Prämissen gerecht zu werden, muss die Agrarforschung Lösungen und **Strategien für angepasste Produktionssysteme** bereitstellen – und das in immer kürzeren Zeitspannen. Der Einsatz von computergestützten Pflanzenwachstumsmodellen, mit deren Hilfe komplexe Anbausysteme regional und überregional, sowie über längere Zeiträume hinweg simuliert und analysierte werden können, hat sich dabei als wertvoll erwiesen. Modellierung und Simulation tragen somit dazu bei Forschung voranzutreiben und gewährleisten, dass Anbausysteme vielschichtiger und intensiver untersucht werden können.

Wie Intercropping Systeme gestaltet werden müssen und welche Probleme dabei auftauchen, welche Datengrundlage für eine Modellierung benötigt wird und welche systemimmanenten Interaktionen berücksichtig werden müssen, sind Gegenstand der vorliegenden Dissertation. Die Dissertation war eingebunden in das Deutsch-Chinesische Graduiertenkolleg „Modellierung von Stoffflüssen und Produktionssystemen für eine nachhaltige Ressourcennutzung in intensiven Acker- und Gemüsebausystemen in der nordchinesischen Tiefebene" und wurde von der Deutschen Forschungsgemeinschaft und dem Chinese Ministry of Education finanziert. Die Datengrundlage basierte auf Feldversuchen, die in China in Form eines Mais/Erdnuss Strip Intercropping, und in Deutschland in Form eines Mais/Weizen und eines Mais/Erbsen Strip Intercropping in den Jahren 2007 bis 2009 durchgeführt wurden.

Die Auswertung der Daten belegte, dass Strip Intercropping auf einem Feldrand-Effekt basiert, Intercropping Modellierung somit auch als Feldrand-Modellierung bezeichnet werden kann. Allerdings gestaltet sich die statistische Auswertung von speziell Strip Intercropping als schwierig, da Intercropping-Versuche aufgrund der zwangsläufig streifenförmigen Anordnung nicht randomisiert werden können. Intercropping bedarf also einer räumlichen Betrachtungsweise, um ertragsrelevante Effekte adäquat abzuschätzen und statistisch abzusichern. Deshalb wurden die Versuche **geostatistisch** ausgewertet und mehrere räumliche Modelle evaluiert und getestet, um die Modellgüte zu verbessern.

Nicht nur die statistische Auswertung von Intercropping ist diffizil, auch die Datengrundlage von Intercropping in China ist lückenhaft. Im Vergleich zu anderen Ländern wie beispielsweise Indien oder Teilen Afrikas, wo Intercropping gängige Praxis ist, scheint die Dokumentation und Erforschung von Intercropping Systemen in China Nachholbedarf zu haben. In einer Literaturstudie (Kapitel 4) wurde deshalb ein erster Versuch unternommen, China in **agro-klimatische Regionen** hinsichtlich ihres Potentials und ihrer Verbreitung von Getreide betonten Intercropping Systemen einzuteilen. Vier unterschiedliche Regionen konnten detektiert werden: der Nordosten, der Norden,

11. Zusammenfassung

der Nordwesten, das Yellow-Huai River Valley und der Südwesten. Die Anbausysteme wechseln von Nord nach Süd von einer Ernte pro Jahr mit einem großen Potential für Intercropping, zu einem Relay Intercropping oder Double Cropping von hauptsächlich Mais und Weizen bis hin zu drei Ernten pro Jahr mit unterschiedlichen Fruchtfolgen, in denen Intercropping entweder integriert ist oder durch eine solche ersetzt wird. Das Kulturarten-Spektrum reicht von Weizen, Mais, Baumwolle, Gründüngung, Soja, Süßkartoffel, Raps, Erdnuss, Millet, Ackerbohne, Buchweizen, Tabak, Sorghum, Reis, Cassava, bis hin zu Knoblauch und einer Vielzahl von verschiedenen Gemüsearten.

In einer zweiten Literaturstudie (Kapitel 5) wurde dargestellt, welche Modelle für Intercropping bereits evaluiert, kalibriert und validiert wurden. Rund 20 verschiedene Modelle in diversen Studien wurden rezensiert, wobei auffiel, dass die **Modellierung von interspezifischer Konkurrenz** in Pflanzengesellschaften wie Mixed Cropping oder Intercropping oftmals auf Konkurrenz um solare Einstrahlung beschränkt wurde. Veränderte mikroklimatische Einflüsse oder Konkurrenz um Bodennährstoffe und –wasser wurden seltener oder gar nicht berücksichtigt. Obwohl die Bandbreite der Modelle relativ groß scheint, verfolgen doch viele Modelle ähnliche oder gleiche Ansätze, wenn es darum geht, Konkurrenz um solare Einstrahlung zu modellieren. Meist bildet ein modifiziertes Beer-Lambert'sches Gesetz die Grundlage.

Exemplarisch für ein prozess-orientiertes Pflanzenwachstumsmodell, welches multiple Anbausysteme und deren Konkurrenz um Sonnenlicht mithilfe des Beer-Lambert'schen Gesetzes simuliert, wurde in Kapitel 9 **APSIM** gewählt. Dieser in der Forschung recht gängige Ansatz wurde mit dem in der vorliegenden Dissertation evaluierten, getesteten und in DSSAT implementierten Beschattungs-Algorithmus verglichen. Mit dem DSSAT Modell war es bislang nicht möglich, Intercropping zu simulieren. Zusätzlich zur Evaluierung des Beschattungs-Algorithmus wurde im Modellierungsansatz mit **DSSAT** ein verändertes Mikroklima in Intercropping verglichen mit Monocropping untersucht.

Es zeigte sich, dass es mit einem modifizierten Beer-Lambert'schen Gesetz nicht möglich war, Strip Intercropping adäquat zu simulieren. Unter der Voraussetzung, dass es im Strip Intercropping einen Gewinner und einen Verlierer gibt, das heißt, dass eine Kulturart mehr Sonnenlicht erhält als im Monocropping und eine andere dafür weniger, ist der Beer-Lambert'sche Ansatz viel versprechend und verwendbar. Die Kompensationsfähigkeit einer Fruchtart kann jedoch nicht simuliert werden, ebenso keine Ertragssteigerung der im System dominanten Fruchtart. Eine Ressourcenteilung kann simuliert werden, allerdings nicht der Umstand, dass Ressourcenverteilung in Strip Intercropping anderen Gesetzen zu unterliegen scheint. Außerdem ist Konkurrenz um solare Einstrahlung zwar

11. Zusammenfassung

ein wichtiger Einflussfaktor in Intercropping, allerdings nicht der einzige – und möglicherweise auch nicht der dominierende.

Im Gegensatz dazu zeigte sich, dass der **Beschattungs-Algorithmus**, der in DSSAT integriert wurde, beide Systeme – Intercropping und Monocropping – simulieren konnte. Der Grad der Beschattung wurde in Prozent berechnet, indem die Einstrahlung im Intercropping proportional zur Einstrahlung im Monocropping erfasst und in Relation zur benachbarten Pflanzenhöhe gesetzt wurde. Statt ein Konkurrenz-Modul in DSSAT zu integrieren, wurde stattdessen der Weather-Input-File gemäß dem Beschattungsgrad modifiziert. Der Beschattungs-Algorithmus kann relativ leicht auf andere Modelle und auf eine Vielzahl von Kulturart-Kombinationen übertragen werden.

Allerdings wurde in diesem Ansatz zusätzlich berücksichtig und getestet, dass Konkurrenz um solare Einstrahlung nicht die einzig bestimmende ist. Der Beschattungs-Algorithmus konnte zwar einen Teil des Ertragszuwachses im Intercropping erklären beziehungsweise simulieren, allerdings erst die Modifizierung des Boden-Stickstoffgehaltes im Intercropping basierend auf der höhren Bodentemperatur und der damit verbundenen höheren Mineralisierungsrate führte zu einer adäquaten Simulation der erzielten Erträge. **Mikroklimatische Einflüsse,** wie eine veränderte Bodentemperatur und eine veränderte Windgeschwindigkeit in Intercropping verglichen mit Monocropping, wirken sich auf Stickstoff-Verfügbarkeit und Stickstoff-Kreislauf im Boden und auf Transpiration und CO_2-Assimilation aus. Der Allokation von Pflanzenwachstumsfaktoren in Raum und Zeit kommt in Intercropping Systemen eine größere Rolle zu als deren absolute Höhe oder Menge. Solche Effekte müssen berücksichtig werden, um die Modellierung von Strip Intercropping weiterhin zu verbessern und Strip Intercropping Systeme zu optimieren.

12 References

Acock, B. (1989): Crop modeling in the USA. Acta Horticulturae 248, pp. 365-371.

Adiku, S.G.K., Carberry, P.S., Rose, C.W., McCown, R.L., and Braddock, R. (1995): A maize (*Zea mays*)-cowpea (*Vigna unguiculata*) intercrop model. H. Sinoquet and P. Cruz (Eds.): Ecophysiology of Tropical Intercropping, Paris: INRA editions, pp. 397-406.

Agyare, W.A., Cloltey, V.A., Mercer-Quarshie, H., and Kambiok, J.M. (2006): Maize yield response in a long-term rotation and intercropping systems in the Guinea savannah zone of Northern Ghana. Journal of Agronomy 5, pp. 232-238.

Aikman, D.P. and Benjamin, L.R. (1994): A model for plant and crop growth, allowing for competition for light by the use of potential and restricted projected crown zone areas. Annals of Botany 73, pp. 185-194.

Alternative Agriculture News (1994): Contour strip intercropping can reduce erosion. AANews, March 1994, Greenbelt, USA.

Altieri, M.A. (2000): Developing sustainable agricultural systems for small farmers in Latin America. Natural Resources Forum 24, pp. 97-105.

Andersen, M.K., Hauggaard-Nielsen, H., Weiner, J., and Jensen, E.S. (2007): Competitive dynamics in two- and three-component intercrops. Journal of Applied Ecology 44, pp. 545-551.

Andow, D.A. (1991): Vegetational diversity and arthropod population response. Annual Review of Entomology 36, pp. 561-586.

Andrighetto, I., Mosca, G., Cozzi, G., and Berzaghi, P. (1992): Maize-soybean intercropping: effect of different variety and sowing density of the legume on forage yield and silage quality. Journal of Agronomy and Crop Science 168, pp. 354-360.

Asseng, S., Keating, B.A., Fillery, I.R.P., Gregory, P.J., Bowden, J.W., Turner, N.C., Palta, J.A., and Abrecht, D.G. (1998): Performance of the APSIM-wheat model in Western Australia. Field Crops Research 57, pp. 163-179.

Asseng, S., Van Keulen, H., and Stol, W. (2000): Performance and application of the APSIM Nwheat model in the Netherlands. European Journal of Agronomy 12, pp. 37-54.

Atlas of the People's Republic of China (1989), 1st ed., Foreign Languages Press, China Cartographic Pub. House, Beijing.

Ayisi, K.K., Putnam, D.H., Vance, C.P., Russelle, M.P., and Allan, D.L. (1997): Strip intercropping and nitrogen effects on seed, oil and protein yields of canola and soybean. Agron. J. 89, pp. 23-29.

Bailey, R.A., Azais, J.M., and Monod, H. (1995): Are neighbour methods preferable to analysis of variance for completely systematic designs? Silly designs are silly! Biometrika 82, pp. 655-659.

Ball, D.A. and Shaffer, M.J. (1993): Simulating resource competition in multispecies agricultural plant communities. Weed Research 33, pp. 299-310.

12. References

Banik, P., Midya, A., Sarkar, B.K., and Ghose, S.S. (2006): Wheat and chickpea intercropping systems in an additive series experiment: Advantages and weed smothering. Europ. J. Agronomy 24, pp. 325-332.

Bartlett, M.S. (1978): Nearest neighbor models in the analysis of field experiments (with Discussion). Journal of Royal Statistical Society, Series B (40), pp. 147-174.

Bauer, S. (2002): Modeling competition with the field-of-neighbourhood approach – from individual interactions to population dynamics of plants. Diss. an der Philipps-Universität Marburg, Leipzig.

Baumann, D.T., Bastiaans, L., Goudriaan, J., Van Laar, H.H., and Kropff, M.J. (2002): Analysing crop yield and plant quality in an intercropping system using an eco-physiological model for interplant competition. Agricultural System 73, pp. 173-203.

Beets, W.C. (1982): Multiple cropping and tropical farming systems. Boulder, Colo.: Westview Pr [u.a.].

Berger, U., Piou, C., Schiffers, K., and Grimm, V. (2008): Competition among plants: Concepts, individual-based modelling approaches, and a proposal for a future research strategy. Perspectives in Plant Ecology, Evolution and Systematics 9, pp. 121-135.

Berntsen, J., Haugaard-Nielsen, H., Olesen, H., Petersen, B.M., Jensen, E.S., and Thomsen, A. (2004): Modelling dry matter production and resource use in intercrops of pea and barley. Field Crops Research 88, pp. 59-73.

Binder, J., Graeff-Hönninger, S., Link, J., Claupein, W., Liu, M., Dai, M., and Wang, P. (2008): Model approach to quantify production potentials of summer maize and spring maize in the North China Plain. Agronomy Journal 100, pp. 862-873.

Binder, J., Graeff-Hönninger, S., Claupein, W., Liu, M., Dai, M., and Wang, P. (2007): An empirical evaluation of yield performance and water saving strategies in a winter wheat – summer maize double cropping system in the North China Plain. Pflanzenbauwissenschaften 11, pp. 1-11.

Blauert, J. and Zadek, S. (1998): Mediating Sustainability. Kumarian Press, Connecticut.

Böning-Zilkens, M.J. (2004): Comparative appraisal of different agronomic strategies in a winter wheat – summer maize double cropping system in the North China Plain with regard to their contribution to sustainability. Berichte aus der Agrarwissenschaft D 100 (Diss. Universität Hohenheim), Aachen.

Brisson, N., Bussiére, F., Ozier-Lafontaine, H., Tournebize, R., and Sinoquet, H. (2004): Adaptation of the crop model STICS to intercropping; Theoretical basis and parameterisation. Agronomie 24, pp. 409-421.

Büro für Technikfolgen-Abschätzung beim Deutschen Bundestag (TAB) (Ed.) (2006): Renaissance des Mischanbaus? Ein traditionelles Verfahren im modernen Gewand. TAB-Brief Nr. 29, pp. 23-26.

Cai, C., Shi, Q., Zhang, S., Jiang, J., Wang, Q., Wang, W., and Yang, X. (1996): Technologies for high yield of groundnut in Pingdu County. C.L.L. Gowda, S.N. Nigam, C. Johansen, and C. Renard (Eds.): Achieving high groundnut yields, Proceedings of an international workshop, 25.-29. August 1995, Shandong Peanut Research Institute (SPRI), ICRISAT, India, pp. 213-216.

12. References

Caldwell, R.M. (1995): Simulation models for intercropping systems. H. Sinoquet and P. Cruz (Eds.): Ecophysiology of Tropical Intercropping, Paris: INRA editions, pp. 353-368.

Campbell, G.S. and Norman, J.M. (1998): An introduction to environmental biophysics. 2nd edition, Springer Science+Business Media, New York.

CANOPY (2010): The APSIM canopy module. APSIM Vs 6 documentation files "Program_Files\Apsim6\apsim\canopy\docs\Canopy_science.htm".

Carberry, P.S., Adiku, S.G.K., McCown, R.L., and Keating, B.A. (1996): Application of the APSIM cropping systems model to intercropping systems. O. Ito, C. Johansen, J.J. Adu-Gyamfi, K. Katayama, J.V.D.K. Kumar Rao, and T.J. Rego (Eds.): Dynamics of roots and nitrogen in cropping systems of the semi-arid tropics, Japan International Research Center for Agricultural Sciences, pp. 637-648.

Carberry, P.S., McCown, R.L., Muchow, R.C., Dimes, J.P., Probert, M.E., Poulton, P.L., and Dalgliesh, N.P. (1996): Simulation of a legume ley farming system in northern Australia using the agricultural production systems simulator. Australian Journal of Experimental Agriculture 36, pp. 1037-1048.

Chalk, P.M. (1998): Dynamics of biologically fixed N in legume-cereal rotations: a review. Australian Journal of Agricultural Research 49, pp. 303-316.

Chapman, H.W., Gleason, L.S., and Loomis, W.E. (1954): The carbon dioxide content of field air. Plant Physiol. 29, pp. 500-503.

Chen, C., Fletcher, S.M., Zhang, P., and Carley, D.H. (2009): Competitiveness of peanuts: United States versus China. The University of Georgia, Cooperative Extension, Research Bulletin 430.

Chen, C., Westcott, M., Neill, K., Wichmann, D., and Knox, M. (2004): Row configuration and nitrogen application for barley-pea intercropping in Montana. Agron. J. 96, pp. 1730-1738.

Chen, Y., Zhang, F., Tang, L., Zheng, Y., Li, Y., Christie, P., and Li, L. (2006): Wheat powdery mildew and foliar N concentrations as influenced by N fertilization and belowground interactions with intercropped faba bean. Plant Soil, doi 10.1007/s11104-006-9161-9.

China Meteorological Administration (2007): Weather data. Beijing, 2007.

China Provincial Statistical Yearbooks (2001-2005), All China Data Center, University of Michigan.

China Yearly Macro-Economics (National) (2010), All China Data Center, University of Michigan.

Chinese Academy of Science (1997): Chinese soil survey 1997. Beijing.

Chinese Academy of Science (Ed.) (1999): The national physical atlas of China. China Cartographic Publishing House, Beijing.

Cochran, W.G. and Cox, G.M. (1957): Experimental designs. Wiley, New York.

Cohen, R. (Ed.) (1988): Satisfying Africa's food needs: food production and commercialization in African agriculture. Book series: Carter studies on Africa, Boulder (u.a..): Rienner.

12. References

Connolly, J., Goma, H.C., and Rahim, K. (2001): The information content of indicators in intercropping research. Agriculture, Ecosystems and Environment 87, pp. 191-207.

Corre-Hellou, G., Faure, M., Launay, M., Brisson, N., and Crozat, Y. (2009): Adaption of STICS intercrop model to simulate crop growth and N accumulation in pea-barley intercrops. Field Crops Research 113, pp.72-81.

Cruse, R.M. (1996): Strip intercropping: A CRP conversion option. Conservation Reserve Program: Issues and Options, Iowa State University publication, University Extension, CRP – 17.

Cullis, B.R. and Gleeson, A.C. (1991): Spatial analysis of field experiments - An extension to two dimensions. Biometrics 47, pp. 1449-1460.

Dawo, M.I., Wilkinson, J.M., Sanders, F.E.T., and Pilbeam, D.J. (2007): The yield of fresh and ensiled plant material from intercropped maize (*Zea mays*) and beans (*Phaseolus vulgaris*). Journal of the Science of Food and Agriculture 87, pp. 1391-1399.

Debaeke, P., Caussanel, J.P., Kiniry, J.R., Kafiz, B., and Mondragon, G. (1997): Modelling crop:weed interactions in wheat with ALMANAC. Weed Research 37, pp. 325-341.

Deneke, H. (1931): Über den Einfluß bewegter Luft auf die Kohlensäureassimilation. Jb. wiss. Bot. 74, pp. 1-32.

EarthTrends (2003): Country profiles; Agriculture and food – China. http://earthtrends.wri.org/pdf_library/country_profiles/agr_cou_156.pdf.

Edmondson, R.N. (2005): Past developments and future opportunities in the design and analysis of crop experiments. Journal of Agricultural Science 143, pp. 27-33.

Ellis, E.C. and Wang, S.M. (1997): Sustainable traditional agriculture in the Tai Lake region of China. Agriculture, Ecosystems and Environment 61, pp. 177-193.

Ericsson-Zenith, S. (1992): Process interaction models. These de Doctorat de L'Universite Pierre et Marie Curie, Paris (VI).

Evans, L.T. (1998): Feeding the ten billion; Plants and population growth. Cambridge University Press.

Evans, L.T. and Wardlaw, I.F. (1976): Aspects of the comparative physiology of grain yield in cereals. Adv. Agron. 28, pp. 301-359.

Fang, Q., Yu, Q., Wang, E., Chen, Y., Zhang, G., Wang, J., and Li, L. (2006): Soil nitrate accumulation, leaching and crop nitrogen use as influenced by fertilization and irrigation in an intensive wheat-maize double cropping system in the North China Plain. Plant and Soil 284, pp. 335-350.

FAO (Ed.) (2006): The state of food and agriculture. FAO Agriculture Series 27, Rome.

FAO (Ed.) (2007): Agricultural production system zone code of the PR China. Communication Division, Rome.

Federer, W.T. (1993): Statistical design and analysis for intercropping experiments, volume I: two crops. Springer, Berlin.

12. References

Federer, W.T. (1998): Statistical design and analysis for intercropping experiments, volume II: three or more crops. Springer, Berlin.

Fernández, C., Acosta, F.J., Abellá, G., López, F., and Díaz, M. (2002): Complex edge effect fields as additive processes in patches of ecological systems. Ecological Modelling 149, pp. 273-283.

Field Crops Research Special Issue (1993): Intercropping – bases of productivity, vol. 34, NOS. 3, 4.

Fisher, R.A. (1925): Statistical methods for research workers. 1st ed. Oliver and Boyd, Edinburgh.

Friis, E., Jensen, J., Mikkelsen, S.A. (1987): Predicting the date of harvest of vining peas by means of air and soil temperature sums and node counting. Field Crops Research 16, pp. 33-42.

Fukai, S. and Trenbath, B.R. (1993): Processes determining intercrop productivity and yields of component crops. Field Crops Research 34, pp. 247-271.

Gai, S., Wang, C., Yu, S., Zhang, J., Zuo, X., Wan., S., Tao., S., Wang, C., and Qiu, R. (1996): Present situation and prospects for groundnut production in China. C.L.L. Gowda, S.N. Nigam, C. Johansen, and C. Renard (Eds.): Achieving high groundnut yields, Proceedings of an international workshop, 25.-29. August 1995, Shandong Peanut Research Institute (SPRI), ICRISAT, India, pp. 17-26.

Gale, F. (2002): China at a glance; a statistical overview of China's food and agriculture. Economic Research Service/USDA (Ed.): China's food and agriculture: Issues for the 21st Century / AIB-775, pp. 5-9.

Gang, Y. (2004): Peanut production and utilization in the People's Republic of China. R.E. Rhoades (Ed.): Peanut in local and global food systems series report no. 4, University of Georgia.

Gao, Y., Duan, A., Sun, J., Li, F., Liu, Z., Liu, H., and Liu, Z. (2009): Crop coefficient and water-use efficiency of winter wheat/spring maize strip intercropping. Field Crops Research 111, pp. 65-73.

García-Barrios, L., Mayer-Foulkes, D., Franco, M., Urquijo-Vásquez, G., and Franco-Pérez, J. (2001): Development and validation of a spatially explicit individual-based mixed crop growth model. Bulletin of Mathematical Biology 63, pp. 507-526.

Garcia Y Garcia, A., Hoogenboom, G., Guerra, L.C., Paz, J.O., and Fraisse, C.W. (2006): Analysis of the inter-annual variation of peanut yield in Georgia using a dynamic crop simulation model. Transactions of the ASABE 49, pp. 2005-2015.

Ghaffarzadeh, M. (1999): Strip intercropping. Iowa State University publication, University Extension, Pn 1763.

Ghaffarzadeh, M., Préchac, F.G., and Cruse, R.M. (1994): Grain yield response of corn, soybean, and oat grown in a strip intercropping system. American Journal of Alternative Agriculture 9, pp. 171-177.

Ghaley, B.S., Hauggaard-Nielsen, H., Høgh-Jensen, H., and Jensen, E.S. (2005): Intercropping of wheat and pea as influenced by nitrogen fertilization. Nutrient Cycling in Agroecosystems 73, pp. 201-212.

Gilmour, A.R., Cullis, B.R., and Verbyla, A.P. (1997): Accounting for natural and extraneous variation in the analysis of field experiments. Journal of Agricultural, Biological and Environmental Statistics 2, pp. 269-293.

Gleeson, A.C. and Cullis, B.R. (1987): Residual maximum likelihood (REML) estimation of a neighbour model for field experiments. Biometrics 43, pp. 277-288.

Gleeson, A.C. (1997): Spatial analysis. R. Kempton and P.N. Fox. (Eds.). Statistical methods for plant variety evaluation, Chapman & Hall, London, pp. 68-85.

Gomes, E.G., da Silva e Souza, G., and Vivaldi, L.J. (2008): Two-stage interference in experimental design using DEA: An application to intercropping and evidence from randomization theory. Pesquisa Operacional 28, pp. 339-354.

Gong, Y., Lin, P., Chen, J., and Hu, X. (2000): Classical farming systems of China. Journal of Crop Production 3, pp. 11-21.

Gowda, C.L.L., Nigam, S.N., Johansen, C., and Renard, C. (Eds.) (1996): Achieving high groundnut yields. Proceedings of an international workshop, 25.-29. August 1995, Shandong Peanut Research Institute (SPRI), ICRISAT, India.

Grace, J. (1988): Plant response to wind. Agriculture, Ecosystems and Environment 22/23, pp. 71-88.

Grant, R.F. (1992): Simulation of competition among plant populations under different managements and climates. Agron. Abstr., American Society of Agronomy, Madison, Wisconsin, USA, 6.

Grant, R.F. (1994): Simulation of competition between barley (*Hordeum vulgare* L.) and wild oat (*Avena fatua* L.) under different managements and climates. Ecological Modelling 71, pp. 269-287.

Gunes, A., Inal, A., Adak, M.S., Alpaslan, M., Bagci, E.G., Erol, T., and Pilbeam, D.J. (2007): Mineral nutrition of wheat, chickpea and lentil as affected by mixed cropping and soil moisture. Nutrient Cycling in Agroecosystems 78, pp. 83-96.

Guohua, X. and Peel, L.J. (1991): The agriculture of China. Published in conjunction with the Centre for Agricultural Strategy, University of Reading, New York.

Haro, R.J., Dardanelli, J.L., Otegui, M.E., and Collino, D.J. (2008): Seed yield determination of peanut crops under water deficit: soil strength effects on pod set, the source-sink ratio and radiation use efficiency. Field Crops Research 109, pp. 24-33.

Harris, D. and Natarajan, M. (1987): Physiological basis for yield advantage in a sorghum/groundnut intercrop exposed to drought. 2. Plant temperature, water status, and components of yield. Field Crops Research 17, pp. 273-288.

Hauggaard-Nielsen, H., Andersen, M.K., Jørnsgaard, B., and Jensen, E.S. (2006): Density and relative frequency effects on competitive interactions and resource use in pea-barley intercrops. Field Crops Research 95, pp. 256-267.

Haymes, R. and Lee, H.C. (1999): Competition between autumn and spring planted grain intercrops of wheat (*Triticum aestivum*) and field bean (*Vicia faba*). Field Crops Research 62, pp. 167-176.

12. References

Herndl, M. (2008): Use of modeling to characterize phenology and associated traits among wheat cultivars. Dissertation, Fakultät Agrarwissenschaften der Universität Hohenheim.

Hinton, W. (1990): The great reversal: The privatization of China, 1978-1989. Monthly Review Press, U.S..

Holzworth, D. and Huth, N. (2009): Reflection + XML simplifies development of the APSIM generic PLANT model. 18th World IMACS / MODSIM Congress, Cairns, Australia 13-17 July 2009, http://mssanz.org.au/modsim09.

Hoogenboom, G., Wilkens P.W., Tsuji, G.Y. (Eds.) (1999): DSSAT v 3, volume 4. University of Hawaii, Honolulu, Hawaii.

Horwith, B. (1985): A role for intercropping in modern agriculture. Bioscience 35, pp. 286-290.

Ibeawuchi, I.I. (2007): Soil-chemical properties as affected by yam/cassava/landrace legumes intercropping systems in Owerri Ultisols Southeastern Nigeria. International Journal of Soil Science 2, pp. 62-68.

Inal, A., Gunes, A., Zhang, F., and Cakmak, I. (2007): Peanut/maize intercropping induced changes in rhizosphere and nutrient concentrations in shoots. Plant Physiology and Biochemistry 45, pp. 350-356.

Inman-Bamber, N.G., Everingham, Y., and Muchow, R.C. (2001): Modelling water stress response in sugarcane: valiadation and application of the APSIM-Sugarcane model. 10th Australian Agronomy Conference, Hobart, Tasmania.

Innis, D.Q. (1997): Intercropping and the scientific basis of traditional agriculture. Intermediate Technology Publications Ltd., London.

Iragavarapu, T.K. and Randall, G.W. (1996): Border effects on yields in a strip-intercropped soybean, corn, and wheat production system. J. Prod. Agric. 9, pp. 101-107.

IUSS Working Group WRB (2007): World Reference Base for Soil Resources 2006, first update 2007. World Soil Resources Reports No. 103. FAO, Rome.

Jensen, E.S. (1996): Grain yield, symbiotic N_2 fixation and interspecific competition for inorganic N in pea-barley intercrops. Plant and Soil 182, pp. 25-38.

Jensen, E.S. (2006): Intercrop; Intercropping of cereals and grain legumes for increased production, weed control, improved product quality and prevention of N-losses in European organic farming systems. Final Report. QLK5-CT-2002-02352, Risø.

Johansen, C. and Nageswara Rao, R.C. (1996): Maximizing groundnut yield. C.L.L. Gowda, S.N. Nigam, C. Johansen, and C. Renard (Eds.): Achieving high groundnut yields, Proceedings of an international workshop, 25.-29. August 1995, Shandong Peanut Research Institute (SPRI), ICRISAT, India, pp. 117-127.

John, J.A. and Williams, E.R. (1995): Cyclic and computer generated designs. Second edition, Chapman and Hall, London.

Jolliffe, P.A. (1997): Are mixed populations of plant species more productive than pure stands? Acta Oecologica Scandinavica (OIKOS) 80, pp. 595-602.

Jones, J.W. (1998): Model integration and simulation tools. Acta Horticulturae 456, pp. 411-417.

Jones, J.W., Hoogenboom, G., Porter, C.H., Boote, K.J., Batchelor, W.D., Hunt, L.A., Wilkens, P.W., Singh, U., Gijsman, A.T., and Ritchie, J.T. (2003): The DSSAT cropping system model. Europ. J. Agronomy 18, pp. 235-265.

Keating, B.A. and Carberry, P.S. (1993): Resource capture and use in intercropping: solar radiation. Field Crops Research 34, pp. 273-301.

Keating, B.A. and Meinke, H. (1998): Assessing exceptional drought with a cropping systems simulator: a case study for grain production in the north-east Australia. Agricultural Systems 57, pp. 315-332.

Kendall, J. (1996/1997): PFI profile: Paul and Karen Mugge. The practical farmer, quaterly newsletter of practical farmers of Iowa 11 (4).

Kiniry, J.R. and Williams, J.R. (1995): Simulating intercropping with the ALMANAC model. H. Sinoquet and P. Cruz (Eds.): Ecophysiology of Tropical Intercropping, Paris: INRA editions, pp. 387-396.

Kiniry, J.R., Williams, J.R., Gassman, P.W., and Debaeke, P. (1992): General, process-oriented model for two competing plant species. Transactions of the American Society of Agricultural Engineers (ASAE) 35, pp. 801-810.

Knörzer, H., Graeff-Hönninger, S., Guo, B., Wang, P., and Claupein, W. (2009): The rediscovery of intercropping in China: a traditional cropping system for future Chinese agriculture. E. Lichtfouse (Ed.): Springer Series: Sustainable Agriculture Reviews 2: Climate Change, Intercropping, Pest Control and Beneficial Microorganisms, Springer Science+Business Media, Berlin, pp. 13-44.

Knörzer, H., Graeff-Hönninger, S., Müller, B.U., Piepho, H.-P., and Claupein, W. (2010): A modeling approach to simulate effects of intercropping and interspecific competition in arable crops. International Journal of Information Systems and Social Change 1, pp. 44-65.

Knörzer, H., Grözinger, H., Graeff-Hönninger, S., Hartung, K., Piepho, H.-P., and Claupein, W. (2010): Integrating a simple intercropping algorithm into CERES-wheat and CERES-maize with particular regard to a changing microclimate within a relay-intercropping system. Field Crops Research (in press).

Knörzer, H., Müller, B.U., Guo, B., Graeff-Hönninger, S., Piepho, H.-P., Wang, P., and Claupein, W. (2010b): Extension and evaluation of intercropping field trials using spatial models. Agronomy Journal 102, pp. 1023-1031.

Koeller, KH. and Linke, C. (2001): Erfolgreicher Ackerbau ohne Pflug: Wissenschaftliche Ergebnisse – Praktische Erfahrungen. 2. neu überarb. u. erw. Aufl., Frankfurt a.M., Germany: DLG.

Kropff, M.J. and Van Laar, H.H. (1993): Modelling crop-weed interactions. CAB International, in association with the International Rice Research Institute.

Kropff, M.J. and Spitters, C.J.T. (1992): An eco-physiological model for interspecific competition, applied to the influence of *Chenopodium album* L. on sugar beet. I. Model description and parameterization. Weed Research 32, pp. 437-450.

12. References

Lei, C. (2005): Statement. Promoting environmentally-friendly agricultural production in China, resource management for sustainable intensive agriculture systems, International Conference in Beijing, China, April 5-7, 2004, Ecological Book Series – 1, Ecological Research for Sustaining the Environment in China, ed. by UNESCO, Beijing, pp. 18-21.

LEL Schwäbisch Gmünd, LTZ Augustenberg, LVVG Aulendorf, LSZ Boxberg, LVG Heidelberg, and HuL Marbach (2007): Stammdatenblätter Landwirtschaft „Nährstoffvergleich Feld-Stall", Tabelle 5a.

Leopold Center (1995) (Ed.): Potential economic, environmental benefits of narrow strip intercropping. Leopold Center Progress Reports 4, pp. 14-19.

Lesoing, G.W. and Francis, C.A. (1999): Strip intercropping effects on yield and yield components of corn, grain sorghum, and soybean. Agron. J. 91, pp. 807–813.

Li, L., Li, S.M., Sun, J.H., Zhou, L.L., Bao, X.G., Zhang, H.G., and Zhang, F. (2007): Diversity enhances agricultural productivity via rhizosphere phosphorus facilitation on phosphorus-deficient soils. PNAS 104, pp. 11192-11196.

Li, L., Sun, J., Zhang, F., Guo, T., Bao, X., Smith, F.A., and Smith, S.E. (2006): Root distribution and interactions between intercropped species. Oecologia 147, pp. 280-290.

Li, L., Sun, J., Zhang, F., Li, X., Rengel, Z., and Yang, S. (2001): Wheat/maize or wheat/soybean strip intercropping II. Recovery or compensation of maize and soybean after wheat harvesting. Field Crops Research 71, pp. 173-181.

Li, L., Sun, J., Zhang, F., Li, X., Yang, S., and Rengel, Z. (2001): Wheat/maize or wheat/soybean strip intercropping I. Yield advantage and interspecific interactions on nutrients. Field Crops Research 71, pp. 123-137.

Li, L., Tang, C., Rengel, Z., and Zhang, F. (2003): Chickpea facilitates phosphorus uptake by intercropped wheat from an organic phosphorus source. Plant and Soil 248, pp. 297-303.

Li, L. and Zhang, F. (2006): Physiological mechanism on interspecific facilitation for N, P and Fe utilization in intercropping systems. 18th World Congress of Soil Science, July 9-15, 2006 – Philadelphia, USA, Saturday 15 July 2006, 166-35, papers.

Li, L., Zhang, F., Li, X., Christie, P., Sun, J., Yang, S., and Tang, C. (2003): Interspecific facilitation of nutrient uptake by intercropped maize and faba bean. Nutrient Cycling in Agroecosystems 65, pp. 61-71.

Li, S.M., Li, L., Zhang, F.S., and Tang, C. (2004): Acid phosphatase role in chickpea/maize intercropping. Annals of Botany 94, pp. 297-303.

Li, W. (2001): Agro-ecological farming systems in China. Man and the biosphere series, ed. by J.N.R. Jeffers, v. 26, Taylor & Francis, Paris.

Li, W., Li, L., Sun, J., Guo, T., Zhang, F., Bao, X., Peng, A., and Tang, C. (2005): Effects of intercropping and nitrogen application on nitrate present in the profile of an Orthic Anthrosol in Northwest China. Agriculture, Ecosystems and Environment 105, pp. 483-491.

Li, X. (1990): Recent development of land use in China. GeoJournal 20, pp. 353-357.

12. References

Liang, X. (1996): Status of groundnut cultivation and production in Guangdong. C.L.L. Gowda, S.N. Nigam, C. Johansen, and C. Renard (Eds.): Achieving high groundnut yields, Proceedings of an international workshop, 25.-29. August 1995, Shandong Peanut Research Institute (SPRI), ICRISAT, India, pp. 217-222.

Lin, J.Y. (1998): How did China feed itself in the past? How will China feed itself in the future? Second Distinguished Economist Lecture, Mexico, D.F. CIMMYT.

Liu, C., Yu, J., and Kendy, E. (2001): Groundwater exploitation and its impact on the environment in the North China Plain. IWRA, Water International 26, pp. 265-272.

Lowenberg-De Boer, J., Krause, M., Deuson, R., and Reddy, K.C. (1991): Simulation of yield distributions in millet-cowpea intercropping. Agricultural Systems 36, pp. 471-487.

Lu, C.H., Van Ittersum, M.K., and Rabbinge, R. (2003): Quantitative assessment of resource-use efficient cropping systems: a case study for Ansai in the Loess Plateau of China. European Journal of Agronomy 19, pp. 311-326.

Lu, W. and Kersten, L. (2005): Chinese grain supply and demand in 2010: Regional perspective and policy implications. Landbauforschung Völkenrode 1, pp. 61-68.

Lüth, K.-M. and Preuße, T. (2007): China bewegt die Welt, DLG-Mitteilungen 8, pp. 12-23.

Ma, K.Y., Hao, S.G., Zhao, H.Y., and Kang, L. (2007): Strip intercropping wheat and alfalfa to improve the biological control of the wheat aphid *Macrosiphum avenae* by the mite *Allothrombium ovatum*. Agriculture, Ecosystems and Environment 119, pp. 49-52.

Mandal, B.K., Das, D., Saha, A., and Mohasin, Md. (1996): Yield advantage of wheat (*Triticum aestivum*) and chickpea (*Cicer arietinum*) under different spatial arrangements in intercropping. Indian Journal of Agronomy 41, pp. 17-21.

Marschner, H. (1986): Mineral nutrition of higher plants. London [u.a.]: Acad. Pr..

McCown, R.L., Hammer, G.L., Hargreaves, J.N.G., Holzworth, D.P., and Freebairn, D.M. (1996): APSIM: a novel software system for model development, model testing and simulation in agricultural systems research. Agricultural Systems 50, pp. 255-271.

McCown, R.L., Hammer, G.L., Hargreaves, J.N.G., Holzworth, D.P., and Huth, N.I. (1995): APSIM: an agricultural production system simulation for operational research. Mathematics and Computers in Simulation 39, pp. 225-231.

Meena, O.P., Gaur, B.L., and Singh, P. (2006): Effects of row ration and fertility levels on productivity, economics and nutrient uptake in maize (*Zea mays*) + soybean (*Glycine max*) intercropping system. Indian Journal of Agronomy 51, pp. 178-182.

Meier, U. (1997): Entwicklungsstadien Pflanzen (BBCH Monograph). Biologische Bundesanstalt für Land- und Forstwissenschaft (Ed.), Blackwell Wissenschafts-Verlag, Berlin-Wien [u.a.].

Meinke, H., Carberry, P.S., Cleugh, H.A., Poulton, P.L., and Hargreaves, J.N.G. (2002): Modelling crop growth and yield under the environmental changes induced by windbreaks. 1. Model development and validation. Australian Journal of Experimental Agriculture 42, pp. 875-885.

12. References

Meng, E.C.H., Hu, R., Shi, X., and Zhang, S. (2006): Maize in China; Production systems, constrains, and research priorities. International Maize and Wheat Improvement Center (CIMMYT), Mexico.

Ministry of Agriculture of the People's Republic of China (Ed.) (2004): Report on the state of China's food security. Beijing.

Monsi, M. and Saeki, T. (1953): Über den Lichtfaktor in den Pflanzengesellschaften und seine Bedeutung für die Stoffproduktion. Jpn.J.Bot. 14, pp. 22-52.

Müller, B.U., Kleinknecht, K., Möhring, J., and Piepho, H.-P. (2010): Comparison of spatial models for sugar beet and barley trials. Crop Science 50, pp. 1-9.

Nelson, R.A., Dimes, J.P., Paningbatan, E.P., and Silburn, D.M. (1998): Erosion/productivity modelling of maize farming in the Philippine uplands part I: Parameterising the agricultural production systems simulator. Agricultural Systems 58, pp. 129-146.

Netting, R.McC. (1993): Smallholders, householders; Farm families and the ecology of intensive, sustainable agriculture. Stanford University Press, California.

Niewienda, A. and Heidt, F.D. (1996): SOMBRERO: A PC-tool to calculate shadows on arbitrarily oriented surfaces. Solar Energy 58, pp. 253-263.

Njoku, S.C., Muoneke, C.O., Okpara, D.A., and Agbo, F.M.O. (2007): Effect of intercropping varieties of sweet potato and okra in an ultisol of southwestern Nigeria. African Journal of Biotechnology 6, pp. 1650-1654.

O'Callaghan, J.R., Maende, C., and Wyseure, G.L.C. (1994): Modelling the intercropping of maize and beans in Kenya. Computers and Electronics in Agriculture 11, pp. 351-365.

Ozier-Lafontaine, H., Bruckler, L., Lafolie, F., and Cabidoche, Y.M. (1995): Modelling root competition for water in mixed crops: a basic approach. H. Sinoquet and P. Cruz (Eds.): Ecophysiology of Tropical Intercropping, Paris: INRA editions, pp. 189-187.

Ozier-Lafontaine, H., Lafolie, F., Bruckler, L., Tournebize, R., and Mollier, A. (1998): Modelling competition for water in intercrops: theory and comparison with field experiments. Plant and Soil 204, pp. 183-201.

Park, S.E., Benjamin, L., and Watkinson, A.R. (2003): The theory and application of plant competition models: an agronomic perspective. Annals of Botany 92, pp. 741-748.

Piepho, H.P. (1995): Implications of a simple competition model for the stability of an intercropping system. Ecological Modelling 80, pp. 251-256.

Piepho, H.-P. (2004): An algorithm for a letter-based representation of all-pairwise comparisons. Journal of Computational and Graphical Statistics 13, pp. 456–466.

Piepho, H.-P., Büchse, A., and Richter, C. (2004): A mixed modelling approach for randomized experiments with repeated measures. Journal Agronomy and Crop Sciences 190, pp. 230-247.

Piepho, H.-P., Richter, C., and Williams, E.R. (2008): Nearest neighbour adjustment and linear variance models in plant breeding trials. Biometrical Journal 50, pp. 164-189.

Piepho, H.-P. and Williams, E.R. (2010): Two-dimensional linear variance structures for plant breeding trials. Plant Breeding 129, pp.1-8.

Prabhakar, S.V.R.K. (2007): Review of the implementation status of the outcomes of the world summit on sustainable development – an Asia-Pacific perspective; climate change implications for sustainable development: need for holistic and inclusive policies in agriculture, land, rural development, desertification, and drought, paper for the regional implementation meeting for Asia and the Pacific for the sixteenth session of the Commission on Sustainable Development, United Nations Economic and Social Commission for Asia and the Pacific in collaboration with FAO Regional Office for Asia and the Pacific, UNCAPSA, UNCCD Asia Regional Coordinating Unit and UNEP Regional Office for Asia and the Pacific, Jakarta.

Pridham, J.C. and Entz, M.H. (2008): Intercropping spring wheat with cereal grains, legumes, and oilseed fails to improve productivity under organic Management. Agronomy Journal 100, pp. 1436-1442.

Probert, M.E., Keating, B.A., Thompson, J.P., and Parton, W.J. (1995): Modelling water, nitrogen, and crop yield for a long-term fallow management experiment. Aust. J. Exp. Agric. 35, pp. 941-950.

Project Proposal (2007): Sub-Project 2.1: Design, modeling and evaluation of improved cropping strategies and multi-level interactions in mixed cropping systems in the North China Plain. First draft, University of Hohenheim, Institute of Crop Production and Grassland Research, (unpublished).

Rao, M.R. and Willey, R.W. (1983): Effects of pigeonpea plant population and row arrangement in sorghum/pigeonpea intercropping. Field Crops Research 7, pp. 203-212.

Raynaud, X. and Leadley, P.W. (2005): Symmetry of belowground competition in a spatially explicit model of nutrient competition. Ecological Modelling 189, pp. 447-453.

Ren, T. (2005): Overview of China's cropping systems. Promoting environmentally-friendly agricultural production in China, resource management for sustainable intensive agriculture systems, International Conference in Beijing, China, April 5-7, 2004, Ecological Book Series – 1, Ecological Research for Sustaining the Environment in China, ed. by UNESCO, Beijing, pp. 96-101.

Reisch, E. and Knecht, G. (1995): Betriebslehre. Stuttgart, Germany: Ulmer.

Ritchie, J.T. (1972): Model for predicting evaporation from row crops with incomplete cover. Water Resour. Res. 8, pp. 1204–1213.

Robertson, M.J., Carberry, P.S., Wright, G.C., and Singh, D.P. (1997): Using models to assess the value of traits of food legumes from a cropping systems perspective. Invited paper for the International Food Legumes Conference, September 1997, Adelaide.

Robertson, M.F., Cawthray, S., Birch, C.J., Bidstrup, R., and Hammer, G.L. (2003): Improving the confidence of dryland maize growers in the northern region by developing low-risk production systems; Solutions for a better environment. Proceedings of the 11th Australian Agronomy Conference Geelong, Victoria.

12. References

Robertson, M., Gaydon, D., Latta, R., Peoples, M., and Swan, A. (2004): Simulating lucerne/crop companion farming systems in Australia. "New directions for a diverse planet", Proceedings of the 4th International Crop Science Congress, 26 Sep – 1 Oct 2004, Brisbane, Australia. Published on CDROM, Web site www.regional.org.au/au/cs.

Robertson, M.J., Whish, J., and Smith, F.P. (2001): Simulating competition between canola and wild radish. ARAB 2001.

Rossiter, D.G. and Riha, S.J. (1999): Modeling plant competition with the GAPS object-oriented dynamic simulation model. Agronomy Journal 91, pp. 773-783.

Russell, G., Jarvis, P.G., and Monteith, J.L. (1989): Absorption of radiation by canopies and stand growth. G. Russell, B. Marshall, and P.G. Jarvis (Eds.): Plant canopies: Their growth, form and function, Cambridge University Press, Cambridge-New York- New Rochelle-Melbourne-Sydney.

Sanders, R. (2000): Prospects for sustainable development in the Chinese countryside; the political economy of Chinese ecological agriculture. Aldershot - Brookfield USA – Singapore - Sydney.

SAS Institute (2009): The SAS System for Windows. Release 9.2. SAS Institute, Cary, NC.

Schabenberger, O. and Gotway, C. (2005): Statistical methods for spatial data analysis. CRC Press, Boca Raton.

Schwarzbach, E. (1984): A new approach in the evaluation of field trials. The determination of the most likely genetic ranking of varieties. Vorträge für Pflanzenzüchtung 6, pp. 249-259.

Sellami, M.H. and Sifaoui, M.S. (1999): Modelling solar radiative transfer inside the oasis; Experimental validation. Journal of Quantitative Spectroscopy & Radiative Transfer 63, pp. 85-96.

Sinoquet, H. and Caldwell, R.M. (1995): Estimation of light capture and partitioning in intercropping systems. H. Sinoquet and P. Cruz (Eds.): Ecophysiology of Tropical Intercropping, Paris: INRA editions, pp. 79-97.

Sinoquet, H., Rakocevic, M., and Varlet-Grancher, C. (2000): Comparison of models for daily light partitioning in multispecies canopies. Agricultural and Forest Meteorology 101, pp. 251-263.

Song, Y.N., Marschner, P., Li, L., Bao, X.G., Sun, J.H., and Zhang, F.S. (2007): Community composition of ammonia-oxidizing bacteria in the rhizosphere of intercropped wheat (*Triticum aestivum* L.), maize (*Zea mays* L.) and faba bean (*Vicia faba* L.). Biol Fertil Soils 44, pp. 307-314.

Song, Z. and Wei, L. (1998): The correlation between windbreak influenced climate and crop yield. International Research Centre, Ottawa, DRC: Library: Documents: Agroforestry Systems in China.

Sonohat, G., Sinoquet, H., Varlet-Grancher, C., Rakocevic, M., Jacquet, A., Simon, J.-C., and Adam, B. (2002): Leaf dispersion and light partitioning in three-dimensionally digitized tall fescue-white clover mixtures. Plant, Cell and Environment 25, pp. 529-538.

Spitters, C.J.T., Toussaint, H.A.J.M., and Goudriaan, J. (1986): Separating the diffuse and direct component of global radiation and its implications for modeling canopy photosynthesis, Part I: Components of incoming radiation. Agricultural and Forest Meteorology 38, pp. 217-229.

12. References

Stilma, E., Keesman, K.J., and Van der Werf, W. (2009): Recruitment and attrition of associated plants under a shading crop canopy: Model selection and calibration. Ecological Modelling 220, pp. 1113-1125.

Su, P., Zhao, A., and Du, M. (2004): Functions of different cultivation modes in oasis agriculture on soil wind erosion control and soil moisture conservation. Chinese Journal of Applied Ecology 15, pp. 1536-1540.

Sullivan, P. (2003): Intercropping principles and production practices. Agronomy Systems Guide, ATTRA 1-800-346-9140, http://attra.ncat.org/attra-pub/PDF/intercrop.pdf.

Thornton, P.K., Dent, J.B., and Caldwell, R.M. (1990): Applications and issues in the modelling of intercropping systems in the tropics. Agriculture, Ecosystems and Environment 31, pp. 133-146.

Tinsley, R.L. (2004): Developing smallholder agriculture – a global perspective. Brussels, Belgium: AgBé Publishing.

Tong, C., Hall, C.A.S., and Wang, H. (2003): Land use change in rice, wheat and maize production in China (1961-1998). Agriculture, Ecosystems and Environment 95, pp. 523-536.

Tsubo, M. and Walker, S. (2002): A model of radiation interception and use by maize-bean intercrop canopy. Agricultural and Forest Meteorology 110, pp. 203-215.

Tsubo, M. and Walker, S. (2004): Shade effects on *Phaseolus vulgaris* L. intercropped with *Zea mays* L. under well-watered conditions. J. Agronomy & Crop Science 190, pp. 168-176.

Tsubo, M., Walker, S., and Mukhala, E. (2001): Comparisons of radiation use efficiency of mono-/inter-cropping systems with different row orientations. Field Crops Research 71, pp. 17-29.

Tsubo, M., Walker, S., and Ogindo, H.O. (2005): A simulation of cereal-legume intercropping systems for semi-arid regions. I. Model development. Field Crops Research 93, pp. 10-22.

Tsuji, G.Y., Uehara, G., and Balas, S. (Eds.) (1994): DSSAT v3. University of Hawaii, Honolulu, Hawaii.

Vandermeer, J. (1989): The ecology of intercropping. Cambridge - New York – New Rochelle – Melbourne – Sydney.

VDLUFA (Ed.) (1991): Methodenbuch Band 1; Die Untersuchung von Böden. 4. Aufl., VDLUFA Verlag, Speyer.

Wang, F., He, Z., Sayre, K., Li, S., Si, Y., Feng, B., and Kong, L. (2009): Wheat cropping systems and technologies in China. Field Crops Research 111, pp. 181-188.

Wen, D., Tang, Y., Zheng, X., and He, Y. (1992): Sustainable and productive agricultural development in China. Agriculture, Ecosystems and Environment 39, pp. 55-70.

Wiles, L.J. and Wilkerson, G.G. (1991): Modeling competition for light between soybean and broadleaf weeds. Agricultural Systems 35, pp. 37-51.

Williams, E.R. (1986): A neighbor model for field experiments. Biometrika 73, pp. 279-287.

12. References

Williams, M.J. (2003): Intercropping Process. United States Patent, no.: US 6,631,585 B1, Oct. 14, 2003.

Williams, P.A. and Gordon, A.M. (1995): Microclimate and soil moisture effects of three intercrops on the rows of a newly-planted intercropped plantation. Agroforestry Systems 29, pp. 285-302.

Wilkinson, G.N., Eckert, S.R., Hancock, T.W., and Mayo, O. (1983): Nearest neighbour analysis of field experiments (with discussion). Journal of Royal Statistical Society B 45, pp. 151-211.

Wilson, J.W. and Wadsworth, R.M. (1958): The effect of wind speed on assimilation rate – a re-assessment. Annals of Botany 22, pp. 285-290.

Wolfinger, R.D. (1996): Heterogenous variance-covariance structures for repeated measures. Journal of Agricultural, Biological, and Environmental Statistics 1, pp. 205-230.

Wubs, A.M., Bastiaans, L., and Bindraban, P.S. (2005): Input Levels and intercropping productivity: exploration by simulation. Plant Research International, note 369 Wageningen, October.

Xi, Y. (Ed.) (1989): Atlas of the People's Republic of China. 1st ed., Foreign Language Press, China Cartographic Pub. House, Beijing.

Xiao, Y., Li, L., and Zhang, F. (2004): Effect of root contact on interspecific competition and N transfer between wheat and fababean using direct and indirect ^{15}N techniques. Plant and Soil 262, pp. 45-54.

Xiaofang, L. (1990): Recent development of land use in China. GeoJournal 20.4, pp. 353-357.

Yamazaki, Y., Kubota, J., Nakawo, M., and Mizuyama, T. (2005): Differences in water budget among wheat, maize and their intercropping field in the Heihe River Basin of Northwest China. AOGS Paper, Asia Oceania Geoscience Society (AOGS) 2nd Annual Meeting 20 to 24 June 2005 in Singapore.

Ye, Y., Li, L., Zhang, F., Sun, J., and Liu, S. (2004): Effect of irrigation on soil NO_3^--N accumulation and leaching in maize/barley intercropping field. Transactions of the Chinese Society of Agricultural engineering 20, pp. 105-109.

Ye, Y., Sun, J., Li, L., and Zhang, F. (2005): Effect of wheat/maize intercropping on plant nitrogen uptake and soil nitrate nitrogen concentration. Transactions of the Chinese Society of Agricultural engineering 21, pp. 33-37.

Yezioro, A. and Shaviv, E. (1994): A design tool for analyzing mutual shading between buildings. Solar Energy 52, pp. 27-37.

Yokozawa, M. and Hara, T. (1992): A canopy photosynthesis model for the dynamics of size structure and self-thinning in plant populations. Annals of Botany 70, pp. 305-316.

Younger, L. (1978): Apparatus and method for sowing second crop in standing crop. United States Patent, no.: 4,084,522, Apr. 18, 1978.

Yu, C. and Li., L. (2007): Amelioration of nitrogen difference method in legume intercropping system. China Scientifical Paper Online, project no.: 20040019035, 1-9.

12. References

Yunlong, C. and Smit, B. (1994): Sustainability in Chinese agriculture: challenge and hope. Agriculture, Ecosystems and Environment 49, pp. 279-288.

Zerner, M.C., Gill, G.S., and Vandeleur, R.K. (2008): Effect of height on the competitive ability of wheat with oats. Agronomy Journal 100, pp. 1729-1734.

Zhang, F. and Li, L. (2003): Using competitive and facilitative interactions in intercropping systems enhances crop productivity and nutrient-use efficiency. Plant and Soil 248, pp. 305-312.

Zhang, F., Shen, J., Li, L., and Liu, X. (2004): An overview of rhizosphere processes related with plant nutrition in major cropping systems in China. Plant and Soil 260, pp. 89-99.

Zhang, L. (2007): Productivity and resource use in cotton and wheat relay intercropping. Chapter 6: Development and validation of SUCROS-Cotton: A mechanistic crop growth simulation model for cotton, applied to Chinese cropping conditions. Doctoral dissertation, Wageningen University, Netherlands.

Zhang, L., Spiertz, J.H.J., Zhang, S., Li, B., and Van der Werf, W. (2008): Nitrogen economy in relay intercropping systems of wheat and cotton. Plant Soil 303, pp. 55-68.

Zhang, X., Tang, F., Wang, B., and Wang, Y. (1996): Research and development of technologies for groundnut/wheat intercropping in Henan province. C.L.L. Gowda, S.N. Nigam, C. Johansen, and C. Renard (Eds.): Achieving high groundnut yields, Proceedings of an international workshop, 25.-29. August 1995, Shandong Peanut Research Institute (SPRI), ICRISAT, India, pp. 203-212.

Zhen, L., Routray, J.K., Zoebisch, M.A., Chen, G., Xie, G., and Cheng, S. (2005): Three dimensions of sustainability of farming practices in the North China Plain; A case study from Ningjin County of Shandong Province, PR China. Agriculture, Ecosystems and Environment 105, pp. 507-522.

Zheng, Y., Zhang, F., and Li, L. (2003): Iron availability as affected by soil moisture in intercropped peanut and maize. Journal of Plant Nutrition 26, pp. 2425-2437.

Zuo, Y., Liu, Y., Zhang, F., and Christie, P. (2004): A study on the improvement iron nutrition of peanut intercropping with maize on nitrogen fixation at early stages of growth of peanut on a calcareous soil. Soil Science and Plant Nutrition 50, pp. 1071-1078.

Acknowledgement / Danksagung

Die Liste derer, denen ich zu tiefem Dank verpflichtet bin, ist lang. Auch wenn unter einer Doktorarbeit nur ein Name steht, ist sie nie das Werk eines einzelnen und nie besser als das Team an Helfern, Betreuern, Mitarbeitern, Kollegen, Coautoren, Freunden und Familie hinter der Kulisse. All denen, die mitgewirkt haben am Werden und am Gelingen dieser Doktorarbeit, möchte ich ganz herzlich danken; und einigen ganz besonders und persönlich.

In erster Linie gilt mein Dank meinem Doktorvater **Prof. Dr. Wilhelm Claupein**; für das Gewähren von sehr vielen Freiheiten und dem Unterstützen von eigenen Ideen. Natürlich geht ein Dankeschön auch an meine beiden Zweitprüfer, **Prof. Dr. Torsten Müller** und **Prof. Dr. Reiner Doluschitz**, die sich bereit erklärt und Zeit genommen haben zum Lesen, Fragen und Prüfen.

Für das Gewähren des Stipendiums und die finanzielle Unterstützung bin ich dem **Chinese Ministry of Education** und vor allem der **Deutschen Forschungsgemeinschaft** zu Dank verpflichtet, die mir darüber hinaus auch ermöglicht hat, an etlichen internationalen Konferenzen und Workshops teilzunehmen, und letztendlich auch meinen Forschungsaufenthalt am CSIRO in Perth, Australien, möglich gemacht hat. All die dort gesammelten Erfahrungen und geknüpften Kontakte stehen nicht in meiner Doktorarbeit geschrieben – oder vielleicht doch?

Für die Hilfe vor Ort (Beijing, China) und für die Betreuung meiner Versuche möchte ich **Prof. Dr. Wang Pu** und meinem Counterpart **Guo Buqing** danken. Darin eingeschlossen sind das unermüdliche Team auf der **Versuchsstation Wuqiao**, das mir trotz Hitze, Staub und viel Arbeit immer wieder gezeigt hat, dass ich im Land des Lächelns bin. Der Projektleitung und ihren Mitarbeitern, **Prof. Dr. Reiner Doluschitz**, **Prof. Dr. Zhang Fusuo**, **Prof. Dr. Liu Xuejun**, **Dr. Tang Aohan**, **Dr. Diana Ebersberger** und **Renate Bayer** sowie sämtlichen **Doktoranden-Kollegen** danke ich für interessante Vorträge und Exkursionen während unserer Blockseminare, für lange und fruchtbare Diskussionen und für die eine oder andere lange Nacht in Beijings Kneipen. Insbesondere sei jedoch Koordinatorin **Dr. Diana Ebersberger** genannt. Liebe Diana, tausend Dank dafür, dass ich stets alle bürokratische Formalitäten auf deinem Schreibtisch abladen durfte, für deine stets offenen Ohren für jegliche Probleme oder auftauchenden Schwierigkeiten, deinen reichen Erfahrungsschatz an hilfreichen China-Tipps, von denen ich gezehrt habe, für tolle Exkursionen und Handouts und Info-Rundmails, vor allem aber für die vielen schönen Stunden mit dir nach getaner Arbeit in China und Deutschland.

Zu einer Doktorarbeit braucht es vielleicht nur fachkundigen Rat, gute Daten, einen Computer und Literatur. Zur Doktorandenzeit braucht es viel, viel mehr. Für das prompte unter die Arme greifen, für die stets offene Tür, für Hilfe, wenn sie nötig war, den heißen Kaffee und das Lächeln am

Acknowledgement / Danksagung

Morgen danke ich stellvertretend für alle, die am **Institut für Kulturpflanzenwissenschaften** (340) für eine tolle Arbeitsatmosphäre gesorgt haben, **Thomas Ruopp**, **Gerdi Frankenberger**, **Anita Kämpf**, **Dagmar Mezger** und natürlich **Andrea Richter**.

All meinen **Coautoren, Korrekturlesern, wissenschaftlichen Ratgebern** und **Kritikern** sei gedankt und gesagt: das Ergebnis liegt vor euch. **Bettina Müller**, ich würde sagen, unsere Zusammenarbeit war von Freundschaft und Erfolg gekrönt. Herrn **Prof. Dr. Hans-Peter Piepho** danke ich für das unermüdliche Mitdenken, Lesen und Verbessern. Und verzeihen sie mir die eine oder andere Überstunde, weil ich wieder einmal eine Deadline voll ausgeschöpft habe.

Stellvertretend für alle Mitarbeiter auf der **Versuchstation Ihinger Hof**, die mir so tatkräftig und zahlreich bei meinen Feldversuchen geholfen und mitgearbeitet haben, danke ich Laborleiter **Martin Zahner**, dem Leiter des Versuchswesens, **Helmut Kärcher**, vor allem aber meinem Versuchstechniker **Herbert Grözinger**. Danke für deine Ruhe und Geduld. Von deiner Erfahrung und Fachkenntnis habe ich mehr als einmal profitieren können, und sie haben mir vieles leichter gemacht. Ich habe das IHO-Team wirklich sehr schätzen gelernt und hatte dank euch eine lehrreiche, abwechslungsreiche und tolle Zeit.

Auch meinen studentischen Mitarbeitern **Lenke Solterbeck**, die mir während der gesamten Zeit eine treue und verlässliche Hilfe war, **Laura Hölz**, **Stefanie Baumann**, **Julia Happel** und **Sarah Melzian** sei für ihren unermüdlichen Einsatz rund um meine Versuche gedankt.

Dr. Michael Robertson, **Dr. Roger Lawes** und **Nat Raisbeck-Brown** vom CSIRO in Perth, Australien, sei gedankt für zwei Monate Forschungsaufenthalt in Australien, für ihre Betreuung, Freundschaft und für APSIM.

Meinen Eltern, die mir die Passion für die Landwirtschaft und die Agrarwissenschaften in die Wiege gelegt haben, ohne die ich nicht promoviert hätte. Auch für eure bedingungslose Unterstützung, für das Hinter-mir-Stehen, das Rücken-Freihalten, das zahlreiche vom Flughafen abholen und für eure Geduld, euch meine Sorgen und Nöte anzuhören.

Last but not least danke ich ganz besonders **Prof. Dr. Simone Graeff-Hönninger**. Ohne dich, deine Hilfe und Unterstützung in jeglicher Hinsicht, wäre diese Doktorarbeit nicht zustande gekommen. Ich hätte mir keine bessere und angenehmere Betreuung wünschen können. Sie war stets kollegial und konstruktiv. Deine Tür stand immer weit offen für mich, und wenn ich verzweifelt und voller Fragezeichen in dein Büro kam, habe ich es danach immer hoch motiviert verlassen. Du hast mir vieles ermöglicht, mich bestärkt, und warst für mich stets der Rückhalt, den man während einer Doktorandenzeit zuweilen dringend benötigt. Danke!

I want morebooks!

Buy your books fast and straightforward online - at one of world's fastest growing online book stores! Environmentally sound due to Print-on-Demand technologies.

Buy your books online at
www.morebooks.shop

Kaufen Sie Ihre Bücher schnell und unkompliziert online – auf einer der am schnellsten wachsenden Buchhandelsplattformen weltweit! Dank Print-On-Demand umwelt- und ressourcenschonend produziert.

Bücher schneller online kaufen
www.morebooks.shop

KS OmniScriptum Publishing
Brivibas gatve 197
LV-1039 Riga, Latvia
Telefax: +371 686 204 55

info@omniscriptum.com
www.omniscriptum.com

Printed by Books on Demand GmbH, Norderstedt / Germany